APPLIED RESEARCH FOR
CORN PRODUCTION
IN INDIANA, 2023

APPLIED RESEARCH FOR CORN PRODUCTION IN INDIANA, 2023

DANIEL QUINN

PURDUE UNIVERSITY PRESS
WEST LAFAYETTE, INDIANA

Cataloging-in-Publication Data on file at the Library of Congress.

978-1-62671-257-7 (paperback)

978-1-62671-258-4 (epdf)

Cover image courtesy of Purdue University Department of Agriculture

CONTENTS

ACKNOWLEDGMENTS

This report entails a detailed summary of applied field research trials for corn production systems in Indiana, conducted under the direction of Dr. Daniel Quinn and the Purdue Corn Agronomy team in the Department of Agronomy at Purdue University. The authors extend many thanks to the Purdue Agronomy Center for Research and Education, the Purdue Agricultural Centers, farmer cooperators, and the many industry collaborators and funding agencies who help provide the necessary resources needed to support this research. Special recongnition is extended to Ana Morales, Riley Seavers, Malena Bartaburu, Erick Oliva, Narciso Zapata, and Jose Vaca who assisted with trial organization, data collection and processing, and the preparation of this report. In addition, the authors also extend thanks to Crystal Paris for report booklet design and visiting scholars and undergraduate students Emely Gramajo, Bailey Temple, Mia Higgins, Caroline Carlton, and Sergio Rubiano who assisted with trial organization, data collection, and scouting. Overall, the combined efforts of various colleagues, professionals, students, and farmers are responsible for the success of this research.

The authors would also like to thank those below for their support in 2023:

Indiana Corn Marketing Council	USDA-NIFA
Corteva Agriscience	John Deere
Pioneer	Netafim
Bayer Crop Science	Copperhead Ag
FMC	Purdue University
Keystone Cooperative	Winfield United
NRCS CIG	Becks Hybrids
BASF	Koch Agronomic Services
Brandt	Pivot Bio
National Science Foundation	Nutrien
The Popcorn Board	NC SARE

SUMMARY OF THE 2023 CORN GROWING SEASON IN INDIANA

In 2023, Indiana produced a state record corn yield average of 203 bushels per acre (bu/ac; Figure 1). The 2023 season was highlighted by strong planting conditions and rapid planting progress to start the season (56% planted as of May 15, 2023, 8 percentage points above the 5-year average). Planting progress and conditions allowed for strong emergence and crop establishment heading into dry conditions in late May and June. For example, the Purdue University research farms in West Lafayette, IN and Butlerville, IN received only 1.0 inch and 2.6 inches of precipitation, respectively, in June. Despite stressful conditions observed across the state in June, disease pressure remained low and minimal pollination issues were observed due to timely July rainfall. In addition, below average temperatures and adequate rainfall in both August and September likely improved grain fill conditions, causing above average kernel weight numbers and leading to higher-than-expected yields across INDIANA.

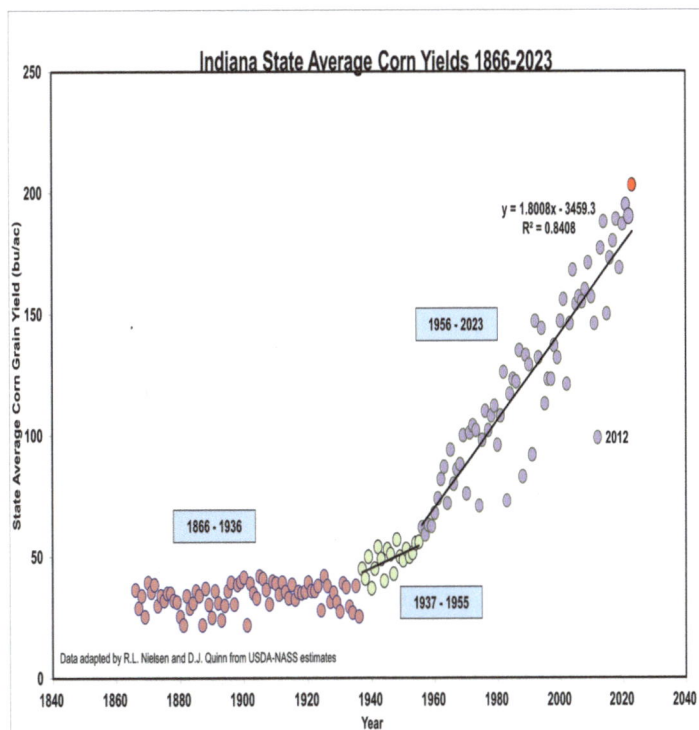

FIGURE 1. Historic state average corn grain yield trends for Indiana (1866–2023).

RANDOMIZATION, REPLICATION, AND STATISTICS: MAKING SENSE OF APPLIED FIELD RESEARCH

Field research trials are an important part of understanding how specific agronomic practices can improve farm productivity. Universities such as Purdue use both research station and on-farm research trials across the state to help drive our recommendations and provide management information for Indiana farmers. However, some of our research practices and conclusions may differ from various private-sector research trials and potentially what you may see on your own farm. For example, you may ask: "Why did they set up the research trial that way?," "What are those letters next to the yield values they are presenting?," and "Why does it seem the university never sees any yield responses from various products?" Therefore, it is important to understand how we approach field research trials, the steps we take to determine our conclusions, and how understanding these approaches can help you better understand and test practices more accurately on your own farm.

Two of the first questions I often ask people when discussing research results are: (1) Do you have a yield monitor in your combine? (2) When traveling across the field during harvest, do those yield values stay the same? The answer I receive 100% of the time is no (if yes, you may need to consider a new monitor), and this is largely due to the variability throughout the field caused by soil type differences, elevation differences, and so on. Therefore, when setting up field research trials we often designate a treatment (e.g., new product) and compare that to a nontreated control (e.g., business as usual). And two of the most important questions we ask after harvest are: (1) Was the yield difference observed truly caused by the product we applied? (2) Or was the yield difference only due to the treated areas being in a more productive part of the field? For example, in Figure 2, if I split a field in half and apply my treatment on one half of the field but don't apply my treatment on the other half of the field, I may find a yield difference of 15 bushels per acre and think to myself, "I should apply this product on all of my acres." However, when you look closer, it is easy to see that the treated area of the field encompassed a larger portion of one soil type, whereas the nontreated area encompassed a larger portion of another soil type. Therefore, it is difficult to determine whether the yield response was due to the product applied or due to the treated area being in a more productive area of the field.

In our university research trials, we approach testing a treatment within a field using randomization, replication (repetition of an experiment in similar conditions), and statistics (Figure 3 and Table 1). For example, compare Figure 2 and Figure 3. Figure 3 highlights how we typically set up one of our research trials using replication and randomization of the treated and nontreated passes to account for field differences. Each of these practices helps us improve the reliability of our conclusions, account for random error (e.g., field variability),

FIGURE 2. Example of a split-field comparison between a nontreated control and a designated treatment.

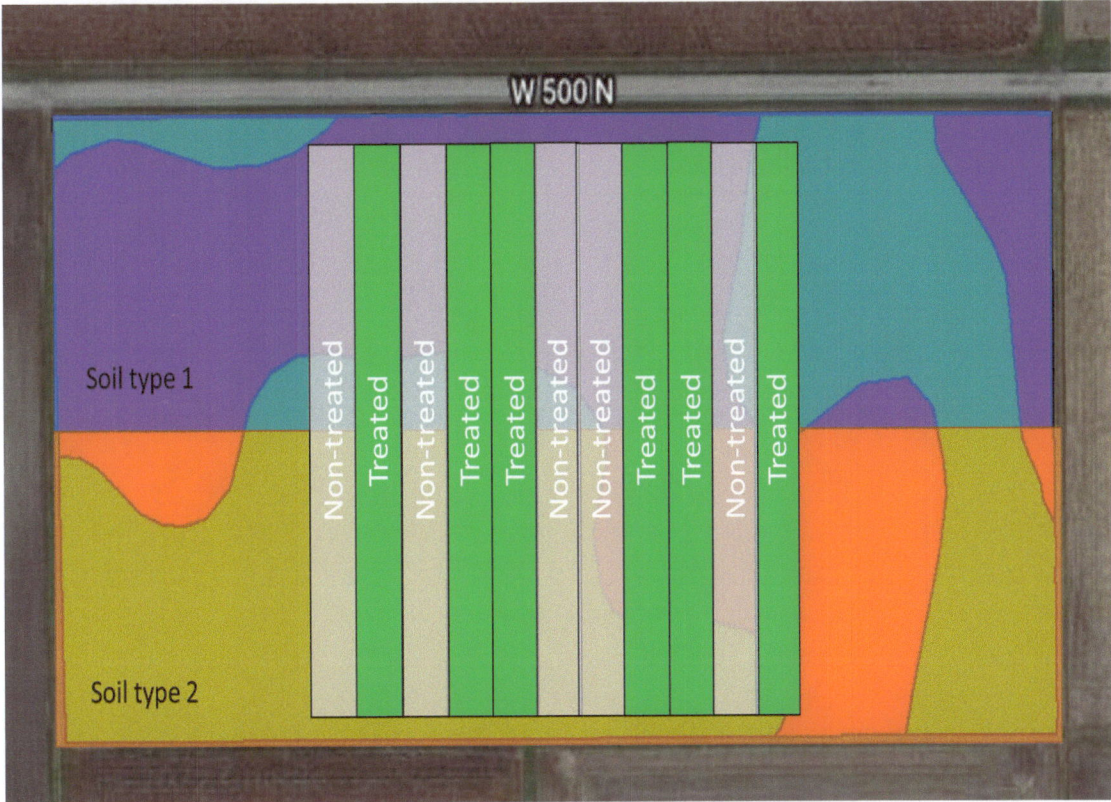

FIGURE 3. Example of a replicated and randomized field research trial comparison between a nontreated control and a designated treatment.

and determine the true causes of yield differences observed. Furthermore, it is also important for us to perform these research trials across multiple locations and multiple years to determine how treatment responses may differ in different fields and different environments. We also use statistical models to help determine our conclusions (Table 1). Using statistics helps us determine if the differences we detect are due to random error or to the treatment we tested. For example, if you have ever seen university data presented (or the data presented in this report), you have probably seen data presented similar to Table 1. At first glance, after we randomized and replicated our treatments (Figure 3), the treated areas seem to have increased corn yield by 4 bushels per acre (Table 1). However, our conclusions suggested no yield differences were observed. Therefore, through the research steps we implemented, it was determined that the yield difference was due to random error (e.g., field variability) and not due to the product or management practice tested. The letters next to the yield values help us highlight where statistical (yield differences due to treatments) differences were observed.

In conclusion, when testing a new product or practice on your own farm, it is important to think about how to design and set up a trial to accurately test the new product or practice. Just because you observe a yield difference doesn't always mean the new product or practice you tested is the reason for this difference. At Purdue, it is our goal to accurately assess new products and practices to determine whether or not these are truly the reason behind observed yield differences. In addition, as you sit in on various meetings, presentations, and examine research results, ask yourself: How did they design and set up this research trial? Did they use randomization, replication, and statistics, and if not, are the yield differences being discussed truly due to the product applied? Over how many different environments and years was this product tested? Understanding and asking these questions can help determine the best products and management practices to implement and improve your operation.

TABLE 1. *Corn grain yield comparisons between the nontreated control and an imposed treatment following a randomized and replicated field research trial.*

TREATMENT	YIELD (BU/AC)
Nontreated	204 a[*]
Treated	208 a

[*] Average yield values that contain the same corresponding letters are not statistically different ($P > 0.1$) from each other.

AGRONOMY CENTER FOR RESEARCH AND EDUCATION (ACRE)

CORN RESPONSE TO INPUT-INTENSIVE MANAGEMENT (ACRE)

Daniel Quinn: Department of Agronomy, Purdue University

Malena Bartaburu: Department of Agronomy, Purdue University

Darcy Telenko: Botany and Plant Pathology, Purdue University

Steven Brand: Botany and Plant Pathology, Purdue University

Rachel Stevens: Purdue Agronomy Center for Research and Education (ACRE)

Study Location: West Lafayette, IN

Soil Type: Chalmers silty clay loam (0–2% slope)

Planting Date: May 6, 2023 | **Harvest Date:** October 4, 2023

Corn Hybrid: Pioneer P1108Q | **Corn Seeding Rate:** 30,000 and 36,000 seeds/ac

Corn Nitrogen (N) Fertilizer Rate and Source: Total N rate across treatments was 180 lbs N/ac, UAN (28-0-0). 40 lbs N/ac applied in a 2x2 starter at planting across all treatments.

Previous Crop: Soybean | **Tillage:** Conventional

Study Replications: 5

RESEARCH TRIAL OVERVIEW:

A field research trial was established at the Purdue Agronomy Center for Research and Education (ACRE) in Tippecanoe County, IN. The research trial examined corn yield response to different management practices and inputs and was utilized to analyze the impact of each one applied individually and in combination. The trial was designed as a randomized complete block design with nine treatments and five replications. Plots measured 15 feet wide (six 30-inch corn rows) by 40 feet long, and the center two rows were harvested with a small-plot combine and adjusted to 15.5% moisture for yield analysis.

RESEARCH TREATMENTS:

1. Control treatment (C) based on Purdue University seed rate (30K seeds/ac) and nitrogen (N) fertilizer rate recommendations
2. C + sub-surface banded starter (2x2) fungicide (flutriafol, Xyway LFR, 15 oz/ac)
3. C + 20% increase in corn seeding rate (36K seeds/ac)
4. C + sulfur fertilizer (5.2 gallons/ac as ammonium thiosulfate (ATS) at V5 sidedress)
5. C + foliar micronutrients (zinc, manganese, and boron applied at the V6 growth stage)
6. C + late-season N application (starter N (2x2) + V5 sidedress N (60% remaining N rate) + V10-12 growth stage sidedress N surface-banded with drop tubes on a sprayer (40% remaining N rate), total N rate remained the same as other treatments).
7. C + foliar fungicide applied at the R1 growth stage (prothioconazole, trifloxystrobin, fluopyram, Delaro Complete, 10 oz/ac)
8. Intensive treatment: all additional inputs (except the Smart KB treatment) and management practices applied together
9. C + foliar fungicide applied at R1 + foliar fertilizer (SmartKB, 2qts/ac)

RESULTS:

TABLE 1. *Corn grain yield, grain moisture, treatment cost, and net profit differences observed from applied treatments in 2023. West Lafayette, IN 2023.*

TREATMENT	GRAIN MOISTURE	GRAIN YIELD	TREATMENT COST§	NET PROFIT§
	--- % ---	-- bu/ac --	-- $/ac --	-- $/ac --
Control	19.2	276.2	253.4	1130.4
C + 2x2 Fungicide	NS*	NS	NS	
C + Increased Seed	NS	NS	NS	NS
C + Sulfur	NS	NS	NS	NS
C + Foliar Micro	NS	NS	NS	NS
C + V10-12 SD N	NS	NS	NS	NS
C + R1 Fungicide	NS	+10.4†	NS	NS
Intensive	NS	+15.4	NS	NS
C + R1 Fungicide + SmartKB	NS	+15.6	NS	NS

* NS, nonsignificant when compared to the control treatment using single degree of freedom contrasts at α = 0.1.

† Mean values that are present within columns indicate statistically significant increase or decrease in the examined variable from the control (α = 0.1).

§ Treatment costs were calculated as the combined cost of corn seed, fertilizer cost, chemical input cost, and application cost. Prices were calculated as an average from various local retailers. Net profit was calculated based on average harvest corn grain cash price ($5.01) + average grain yield – treatment costs.

SUMMARY (TAKE-HOME POINTS):

- The R1 foliar fungicide application, R1 fungicide + SmartKB, increased seeding rate (20% increase), and intensive treatment (all inputs applied) significantly increased corn grain yield in comparison to the control in 2023 at this location (Table 1).
- Despite observed yield responses, no differences in net profit were observed across examined treatments at this location (Table 1).
- Foliar disease presence (e.g., tar spot, gray leaf spot) was significantly reduced by the R1 fungicide application, likely driving the observed yield responses (Table 1).
- No significant yield responses were observed from the 2x2 starter fungicide, foliar micronutrient, delayed side-dress N, and sulfur application at this location in 2023.

CORN RESPONSE TO NITROGEN FERTILIZER APPLICATION TIMING FOLLOWING A RYE COVER CROP (ACRE)

Daniel Quinn: Department of Agronomy, Purdue University

Riley Seavers: Department of Agronomy, Purdue University

Shalamar Armstrong: Department of Agronomy, Purdue University

Rachel Stevens: Purdue Agronomy Center for Research and Education (ACRE)

Evan Bossung: Purdue Agronomy Center for Research and Education (ACRE)

Aaron Kult: Purdue Agronomy Center for Research and Education (ACRE)

Study Location: West Lafayette, IN

Soil Type: Drummer Fine-Silty (0–2% slope), Raub-Brenton complex (0–1% slope)

Planting Date: May 6, 2023 | **Harvest Date:** October 10, 2023

Corn Hybrid: Pioneer P1108Q | **Corn Seeding Rate:** 30,000 seeds/ac

Corn Nitrogen (N) Fertilizer Rate and Source: 200 lbs N per acre, UAN (28-0-0). 40 lbs N/ac were applied in a 2x2 starter at planting across all N timing treatments.

Previous Crop: Soybean | **Tillage:** No-Till

Study Replications: 4

RESEARCH TRIAL OVERVIEW:

This research trial was established at the Purdue Agronomy Center for Research and Education in Tippecanoe County, IN. This research trial assessed corn yield differences when utilizing different N fertilizer application timings following a fall-planted cereal rye cover crop and no rye cover crop. The rye cover crop was fall drill-seeded at a rate of 45 lbs/ac and was chemically terminated with glyphosate two weeks before corn planting. Plots measured 30 feet wide (12, 30-inch corn rows) x 500+ feet long. The center eight rows were harvested with a commercial combine containing a calibrated yield monitor and were adjusted to 15.5% moisture for yield analysis.

RESEARCH TREATMENTS:

1. No cover crop + V5 sidedress N (coulter-inject, UAN 28-0-0)
2. No cover crop + V10 sidedress N (surface-banded, UAN 28-0-0)
3. No cover crop + V5 + V10 sidedress N (60% remaining N at V5 and 40% remaining N at V10)
4. Cereal rye cover crop + V5 sidedress N (coulter-inject, UAN 28-0-0)
5. Cereal rye cover crop + V10 sidedress N (surface-banded, UAN 28-0-0)
6. Cereal rye cover crop + V5 + V10 sidedress N (60% remaining N at V5 and 40% remaining N at V10)

RESULTS:

TABLE 2. *Mean cereal rye cover crop aboveground biomass and nutrient uptake values. Rye cover crop biomass was sampled immediately prior to spring termination. West Lafayette, IN 2023.*

BIOMASS	CARBON	NITROGEN	C:N*
-- lbs/ac --	-- lbs/ac --	-- lbs/ac --	
1248	528.1	43.4	14:1

* Carbon (C) to nitrogen (N) ratio of aboveground biomass at termination.

TABLE 3. *Mean plant population, grain moisture, and grain yield differences observed across cereal rye cover crop presence and nitrogen (N) fertilizer application timing. West Lafayette, IN 2023.*

COVER CROP	N TIMING	V5 PLANT POPULATION	GRAIN MOISTURE	GRAIN YIELD
		-- plants/ac --	-- % --	-- bu/ac --
No Cover	0N	26,656	18.9 d*	210.1 e
	V5	27,281	20.7 c	272.8 a
	V10	27,406	20.6 c	267.2 ab
	V5+V10	27,437	20.5 c	273.1 a
Rye Cover Crop	0N	27,656	20.8 c	196.9 f
	V5	27,563	23.0 a	257.8 bc
	V10	27,562	22.3 b	240.4 d
	V5+V10	27,531	22.3 b	247.3 cd
P–value		0.49	0.01	0.01

* Mean values that do not contain the same corresponding letter are determined statistically different ($P < 0.1$). Columns with mean values that do not contain any letters are determined as no statistical differences between treatments.

SUMMARY (TAKE-HOME POINTS):

- Rye cover crop biomass at termination averaged 1,248 lbs/ac, which contained 528 lbs/ac of carbon, 43 lbs/ac of N, and a C:N ratio of 14:1 (Table 2).
- Across all treatments examined, a rye cover crop reduced corn yield by 20 bu/ac. Yield reductions were likely due to observed N stress and delayed plant growth.
- No statistical differences in grain yield were observed between all examined N timings following no cover crop, and no statistical difference between the V5 and V5+V10 timings were observed when following a rye cover crop (Table 3).
- Grain moisture at harvest was increased by 1.8 points with rye cover crop presence (Table 3).
- Preliminary results suggest an N application at the V5 growth stage is the most beneficial when following a rye cover crop. In addition, delayed sidedress N application did not reduce yield without a rye cover crop, yet did result in a yield reduction when following a rye cover crop.

SHORT-STATURE CORN HYBRID RESPONSE TO ROW SPACING AND PLANT POPULATION (ACRE)

Daniel Quinn: Department of Agronomy, Purdue University

Erick Oliva: Department of Agronomy, Purdue University

Shaun Casteel: Department of Agronomy, Purdue University

Rachel Stevens: Purdue Agronomy Center for Research and Education (ACRE)

Study Location: West Lafayette, IN

Soil Type: Drummer Fine-Silty (0–2% slope)

Planting Date: May 11, 2023 | **Harvest Date:** October 22, 2023

Corn Hybrid: Bayer CropScience RT6203TVXZ, RV6205TVXZ, RW5419KTFZ

Corn Seeding Rate: 34,000; 42,000 and 50,000 seeds/ac

Corn Nitrogen (N) Fertilizer Rate and Source: 200 lbs N/ac, UAN (28-0-0). Nitrogen was applied as a broadcast application prior to trial planting. No starter fertilizer was used in this trial.

Previous Crop: Soybean | **Tillage:** Conventional

Study Replications: 4

RESEARCH TRIAL OVERVIEW:

A field research trial was established at the Purdue Agronomy Center for Research and Education (ACRE) in Tippecanoe County, IN. The research trial examined corn yield response to different short-stature corn hybrids, row spacings, and seeding rates. The trial was designed as a split-plot, randomized complete block design with 18 treatments and four replications. Plots measured 10 feet wide (four 30-inch corn rows and six 20-inch corn rows) by 40 feet long, and the center two rows were harvested with a small-plot combine and adjusted to 15.5% moisture for yield analysis.

RESULTS:

TABLE 4. *Short-stature corn grain moisture and yield in response to row spacing, hybrid, and seeding rate. West Lafayette, IN 2023.*

ROW SPACING	HYBRID	SEEDING RATE	GRAIN MOISTURE	GRAIN YIELD
		-- seeds/ac --	-- % --	-- bu/ac --
20 Inches	RT6203TVXZ	34,000	19.3 a*	287.9 ab
		42,000	19.7 a	297.4 a
		50,000	19.4 a	308.7 a
	RV6205TVXZ	34,000	18.3 b	254.9 c
		42,000	19.0 ab	270.6 b
		50,000	18.4 b	309.5 a
	RW5419KTFZ	34,000	17.4 c	225.1 d
		42,000	17.2 c	245.3 c
		50,000	17.6 c	259.5 c
30 Inches	RT6203TVXZ	34,000	19.8 a	274.6 b
		42,000	19.7 a	299.4 a
		50,000	19.5 a	296.3 a
	RV6205TVXZ	34,000	18.7 b	276.4 b
		42,000	18.8 b	292.2 a
		50,000	18.8 b	280.5 b
	RW5419KTFZ	34,000	17.1 c	248.2 d
		42,000	17.5 c	257.8 c
		50,000	17.7 c	262.9 c
P–value			*0.01*	*0.01*

* Mean values that do not contain the same corresponding letter are determined statistically different ($P < 0.1$). Columns with mean values that do not contain any letters are determined as no statistical differences between treatments.

SUMMARY (TAKE-HOME POINTS):

- Short-stature corn hybrids illustrated a higher tolerance to the higher seeding rate (50,000 plants/ac) and higher yield potential in 20-in rows as compared to 30-in rows (Table 4).
- Short-stature corn hybrid RW5419KTFZ presented lower yields compared to the other hybrids likely due to its low ear height (driven by dry conditions in June), which presented harvest challenges with the combine header.
- When comparing yield differences between the three hybrids selected and across both row spacings and seeding rates, hybrid RT6203TVXZ outyielded hybrids RV6205TVXZ and RW5419KTFZ by an average of 13 and 44 bushels per acre (bu/ac), respectively.
- Preliminary data suggests short-stature corn hybrids can tolerate higher seeding rates in 20-in row spacing as compared to 30-in row spacings. In addition, hybrid yield differences may be dictated by ear height and compatibility with combine specifications. Therefore, short-stature hybrid selection that accounts for average ear height may be necessary to avoid any harvest challenges.

SHORT-STATURE CORN HYBRID RESPONSE TO FUNGICIDE APPLICATION TIMING (ACRE)

Daniel Quinn: Department of Agronomy, Purdue University

Erick Oliva: Department of Agronomy, Purdue University

Steven Brand: Botany and Plant Pathology, Purdue University

Rachel Stevens: Purdue Agronomy Center for Research and Education (ACRE)

Study Location: West Lafayette, IN

Soil Type: Toronto Millbrook complex (0–2% slope)

Planting Date: June 02, 2023 | **Harvest Date:** November 03, 2023

Corn Hybrid: Bayer CropScience RT6203TVXZ, RV6205TVXZ, RV6210TVXZ

Corn Nitrogen (N) Fertilizer Rate and Source: 200 lbs N/ac, UAN (28-0-0). 40 lbs N/ac was applied in a 2x2 starter across all treatments.

Corn Fungicide Product and Rate Used: Delaro Complete, 10 oz/ac, applied at the R1 and R3 growth stages.

Corn Seeding Rate: 32,000 seeds/ac

Previous Crop: Soybean | **Tillage:** Conventional

Study Replications: 5

RESEARCH TRIAL OVERVIEW:

A field research trial was established at the Purdue Agronomy Center for Research and Education (ACRE) in Tippecanoe County, IN. The research trial examined corn yield response to different short-stature corn hybrids and fungicide timings. The trial was designed as a split-plot, randomized complete block design with nine treatments and five replications. Plots measured 15 feet wide (six 30-inch corn rows) by 40 feet long, and the center two rows were harvested with a small-plot combine and adjusted to 15.5% moisture for yield analysis. All plots received a 2x2 starter N application totaling 40 lbs N/ac.

RESULTS:

TABLE 5. *Corn grain moisture and yield in response to short-stature corn hybrids and fungicide timing. West Lafayette, IN 2023.*

HYBRID	FUNGICIDE TIMING	GRAIN MOISTURE	GRAIN YIELD
		-- % --	-- bu/ac --
All Hybrids	No Fungicide	23.3	257.8 b*
	R1	23.2	266.7 ab
	R1+R3	23.7	273.7 a
P–value		*0.34*	*0.04*

* Mean values that do not contain the same corresponding letter are determined statistically different ($P < 0.1$). Columns with mean values that do not contain any letters are determined as no statistical differences between treatments.

SUMMARY (TAKE-HOME POINTS):

- No significant interaction was observed between short-stature corn hybrid type and fungicide application timing response (Table 5). Therefore, data was analyzed across all hybrids present in the research trial.
- Across all hybrids, the R1 + R3 foliar fungicide application statistically outyielded the control (no fungicide application) by 16 bushels per acre. Both tar spot and gray leaf spot were present in this research trial, which likely drove observed yield responses.

SHORT-STATURE CORN HYBRID RESPONSE TO SEEDING RATE (ACRE)

Daniel Quinn: Department of Agronomy, Purdue University

Erick Oliva: Department of Agronomy, Purdue University

Rachel Stevens: Purdue Agronomy Center for Research and Education (ACRE)

Study Location: West Lafayette, IN

Soil Type: Drummer Fine-Silty (0–2% slope)

Planting Date: May 22, 2023 | **Harvest Date:** October 22, 2023

Corn Hybrid: Bayer CropScience RT6203TVXZ, RV6205TVXZ, RV6210TVXZ

Corn Nitrogen (N) Fertilizer Rate and Source: 180 lbs N/ac, UAN (28-0-0). 40 lbs N/ac was applied in a 2x2 starter at planting across all treatments.

Corn Seeding Rate: 32,000; 38,000; 44,000 seeds/ac

Previous Crop: Soybean | **Tillage:** Conventional

Study Replications: 4

RESEARCH TRIAL OVERVIEW:

A field research trial was established at the Purdue Agronomy Center for Research and Education (ACRE) in Tippecanoe County, IN. The research trial examined corn yield response to different short-stature corn hybrids and seeding rates. The trial was designed as a split-plot, randomized complete block design with nine treatments and four replications. Plots measured 15 feet wide (six 30-inch corn rows) by 40 feet long, and the center two rows were harvested with a small-plot combine and adjusted to 15.5% moisture for yield analysis. All plots received a 2x2 starter N application totaling 40 lbs N/ac.

RESULTS:

TABLE 6. *Short-stature corn grain moisture and yield in response to hybrid and seeding rate. West Lafayette, IN 2023.*

HYBRID	SEEDING RATE	GRAIN MOISTURE	GRAIN YIELD
	--seeds/ac--	-- % --	-- bu/ac --
RT6203TVXZ	32,000	21.4 c*	288.5 b
	38,000	21.6 bc	299.1 a
	44,000	21.4 cd	301.4 a
RV6205TVXZ	32,000	20.9 e	276.2 c
	38,000	21.0 de	285.3 b
	44,000	20.9 e	287.7 b
RV6210TVXZ	32,000	22.1 a	279.5 c
	38,000	21.9 ab	288.1 b
	44,000	21.5 c	287.9 b
P-value		0.01	0.01

* Mean values that do not contain the same corresponding letter are determined statistically different ($P < 0.1$). Columns with mean values that do not contain any letters are determined as no statistical differences between treatments.

SUMMARY (TAKE-HOME POINTS):

- Across all short-stature hybrid examined, the greatest grain yields occurred at the 38K and 44K seeding rates.
- Across all seeding rates examined, when comparing yield differences between the 3 hybrids, hybrid RT6203TVXZ outyielded hybrid RV6205TVXZ and hybrid RV6210TVXZ by an average of 13 and 11 bushels per acre (bu/ac), respectively.

COMPARISON OF SHORT- AND TALL-STATURE CORN HYBRIDS TO NITROGEN FERTILIZER AND SEEDING RATES (ACRE)

Daniel Quinn: Department of Agronomy, Purdue University

Erick Oliva: Department of Agronomy, Purdue University

Rachel Stevens: Purdue Agronomy Center for Research and Education (ACRE)

Study Location: West Lafayette, IN

Soil Type: Drummer Fine-Silty (0–2% slope)

Planting Date: May 11, 2023 | **Harvest Date:** October 22, 2023

Corn Hybrid: Bayer Crop Science RT6203TVXZ, RV6205TVXZ, DKC61-41RIB, DKC62-70RIB

Corn Seeding Rate: 32,000; 38,000 and 44,000 seeds/ac

Corn Nitrogen (N) Fertilizer Rate and Source: 160, 200, 240 and 280 lbs. N/ac, UAN (28-0-0) applied at the V5 growth stage. All plots received a 2x2 starter N application totaling 40 lbs N/ac.

Previous Crop: Soybean | **Tillage:** Conventional

Study Replications: 4

RESEARCH TRIAL OVERVIEW:

A field research trial was established at the Purdue Agronomy Center for Research and Education (ACRE) in Tippecanoe County, IN. The research trial examined corn yield response to different short and full-size corn hybrids, nitrogen, and seeding rates. The trial was designed as a split-plot, randomized complete block design with 48 treatments and four replications. Plots measured 15 feet wide (six 30-inch corn rows) by 40 feet long, and the center two rows were harvested with a small-plot combine and adjusted to 15.5% moisture for yield analysis.

RESULTS:

TABLE 7. *Short-stature and tall-stature corn grain moisture and yield in response to hybrid and seeding rate. West Lafayette, IN 2023.*

TARGETED SEED RATE	HYBRID (TYPE)	GRAIN YIELD
		-- bu/ac --
32,000 seeds/ac	RT6203TVX2 (Short)	278.3 c*
	RV6205TVX4 (Short)	274.3 c
	DKC61-41RIB (Tall)	292.2 b
	DKC62-70RIB (Tall)	304.1 a
38,000 seeds/ac	RT6203TVX2 (Short)	288.4 c
	RV6205TVX4 (Short)	284.3 c
	DKC61-41RIB (Tall)	302.8 b
	DKC62-70RIB (Tall)	311.7 a
44,000 seeds/ac	RT6203TVX2 (Short)	297.5 b
	RV6205TVX4 (Short)	281.9 c
	DKC61-41RIB (Tall)	309.8 a
	DKC62-70RIB (Tall)	308.9 a

* Mean values that do not contain the same corresponding letter are determined statistically different ($P < 0.1$). Columns with mean values that do not contain any letters are determined as no statistical differences between treatments.

TABLE 8. *Short-stature and tall-stature corn grain moisture and yield in response to hybrid and nitrogen rate. West Lafayette, IN 2023.*

HYBRID	NITROGEN RATE	GRAIN MOISTURE	GRAIN YIELD
	--lbs N/ac--	-- % --	-- bu/ac --
RT6203TVXZ	160	19.7	289.6*
	200	19.6	286.7
	240	19.6	283.5
	280	20.0	290.1
RV6205TVXZ	160	19.9	283.2
	200	19.9	282.7
	240	19.9	278.3
	280	19.6	276.4
DKC61-41RIB	160	19.0	303.2
	200	18.8	297.5
	240	18.7	300.2
	280	18.7	305.2
DKC62-70RIB	160	19.9	305.4
	200	20.0	305.6
	240	20.0	309.4
	280	20.0	311.1

* Mean values that do not contain the same corresponding letter are determined statistically different ($P < 0.1$). Columns with mean values that do not contain any letters are determined as no statistical differences between treatments.

SUMMARY (TAKE-HOME POINTS):

- A significant hybrid x seeding rate interaction was observed in this research trial, which suggests that the different hybrids examined had different optimum seeding rates in this year and environment (Table 7).
- Across all hybrids examined, the hybrids did not differ in their response to applied total N fertilizer rates (Table 8). Results suggest optimum N fertilizer rates do not differ between short- and tall-stature hybrids.

CORN RESPONSE TO INTENSIVE MANAGEMENT WITH IRRIGATION AND FERTIGATION (ACRE)

Daniel Quinn: Department of Agronomy, Purdue University
Jose Vaca: Department of Agronomy, Purdue University
Laura Bowling: Department of Agronomy, Purdue University
Shaun Casteel: Department of Agronomy, Purdue University
Rachel Stevens: Purdue Agronomy Center for Research and Education (ACRE)
Evan Bossung: Purdue Agronomy Center for Research and Education (ACRE)
Aaron Kult: Purdue Agronomy Center for Research and Education (ACRE)

Study Location: West Lafayette, IN
Soil Type: Chalmers silty clay loam (0–2% slope), Toronto-Millbrook complex (0–2% slope)
Planting Date: May 5, 2023 | **Harvest Date:** October 23, 2023
Corn Hybrid: Pioneer P1108Q | **Corn Seeding Rate:** 32,000 and 38,000 seeds/ac
Corn Nitrogen (N) Fertilizer Rate and Source: 180 lbs N/ac, UAN (28-0-0)
Previous Crop: Soybean | **Tillage:** Conventional
Study Replications: 4

RESEARCH TRIAL OVERVIEW:

This research trial was established at the Agronomy Center for Research and Education (ACRE) in Tippecanoe County, IN. This research trial examined corn grain yield differences under conventional and intensive management with and without the use of recycled drainage water for application of irrigation and fertigation using surface-drip lines. The experimental design of this trial was a randomized complete block design with four replications. All plots measured 60 feet wide (24, 30-inch corn rows) x 65 feet long. All plots received a 2x2 starter fertilizer application of nitrogen at 40 lbs N/ac. The center four rows of each plot were harvested with a small-plot combine and adjusted to 15.5% moisture for yield analysis.

RESEARCH TREATMENTS:

1. Control treatment (C) based on Purdue University seed rate (32K seeds/ac) and nitrogen (N) fertilizer rate recommendations.

2. C + Irrigation. Irrigation water was applied through a surface-drip application. Irrigation applications were decided daily based on a checkbook and soil moisture levels collected from active soil moisture sensors placed at various depths.

3. C + Fertigation* consisted of 20 lbs N/ac applied as UAN (28-0-0) and an additional 2 lbs S/ac applied as ATS (12-0-0-26S) injected through the surface drip lines with irrigation application at the V12, R1, and R2 growth stages.

4. Intensive Management (IM). Seeding rate of 38,000 seeds/ac, multiple in-season N fertilizer application [starter N (2x2) + V5 sidedress N (60% remaining N rate) + V10-12 growth stage sidedress N surface-banded with drop tubes on a sprayer (40% remaining N rate), total N rate remained the same as other treatments], sulfur fertilizer (5.2 gallons/ac as ammonium thiosulfate (ATS) at V5 sidedress), and foliar fungicide at the R1 growth stage (mefentrifluconazole, pyraclostrobin, Veltyma, 10 oz/ac).

5. Intensive Management (IM) + Irrigation. Treatment contained all intensive applications above combined with irrigation presence.

6. Intensive Management (IM) + Fertigation.* Fertigation consisted of 20 lbs N/ac applied as UAN (28-0-0) and an additional 2 lbs S/ac applied as ATS (12-0-0-26S) injected through the surface drip lines with irrigation application at the V12, R1, and R2 growth stages. Additionally, foliar fungicide at the R1 growth stage was applied (mefentrifluconazole, pyraclostrobin, Veltyma, 10 oz/ac).

*All treatments that included fertigation received the same total amount of water as the "irrigation water only" treatments.

RESULTS:

TABLE 9. *Corn grain moisture and grain yield differences observed from applied treatments in 2023. West Lafayette, IN 2023.*

TREATMENT	GRAIN MOISTURE	GRAIN YIELD
	--- % ---	-- bu/ac --
Control (C)	17.83 ab*	267.5 b
C + Irrigation	17.80 ab	281.9 a
C + Fertigation	17.83 ab	277.5 ab
Intensive (IM)	17.57 b	277.2 ab
IM + Irrigation	17.91 a	283.6 a
IM + Fertigation	17.72 ab	285.5 a
P-value	0.56	0.02

* Mean values that do not contain the same corresponding letter are determined statistically different ($P < 0.1$). Columns with mean values that do not contain any letters are determined as no statistical differences between treatments.

SUMMARY (TAKE-HOME POINTS):

- Irrigation, intensive management, intensive management + irrigation, and intensive management + fertigation all increased corn yield beyond the control (Table 9).
- Overall, the data suggests the inclusion of irrigation water resulted in the greatest influence in corn yield beyond additional management factors (e.g., higher seed rate, sulfur, fungicide, etc.).
- Preliminary results suggest recycled drainage water applied through surface-drip lines has the potential to increase corn yield by supplementing crop water needs during the season. However, results will be repeated in the upcoming years to capture different environmental conditions and will also include a transition to subsurface drip line for further evaluation.

CORN RESPONSE TO VARIOUS IN-FURROW AND SEED-APPLIED NUTRIENT PRODUCTS (ACRE)

Daniel Quinn: Department of Agronomy, Purdue University
Rachel Stevens: Purdue Agronomy Center for Research and Education (ACRE)

Study Location: West Lafayette, IN
Soil Type: Drummer Fine-Silty (0–2% slope)
Planting Date: May 5, 2023 | **Harvest Date:** October 10, 2023
Corn Hybrid: Pioneer P1108Q | **Corn Seeding Rate:** 32,000 seeds/ac
Corn Nitrogen (N) Fertilizer Rate and Source: 180 lbs N/ac, UAN (28-0-0).
Previous Crop: Soybean | **Tillage:** Conventional
Study Replications: 4

RESEARCH TRIAL OVERVIEW:

This research trial was established at the Agronomy Center for Research and Education (ACRE) in Tippecanoe County, IN. This research trial examined corn grain yield differences to in-furrow and seed-applied nutrition products from Brandt. The experimental design of this trial was a randomized complete block design with four replications. All plots measured 15 feet wide (six 30-inch corn rows) x 40 feet long. The center two rows of each plot were harvested with a small-plot combine and adjusted to 15.5% moisture for yield analysis.

RESEARCH TREATMENTS:

1. Control treatment (C) with no in-furrow starter fertilizer product applied.
2. Ammonium Polyphosphate (APP; 10-34-0) applied in-furrow at a rate of 5 gallons per acre.
3. EnzUP Phosphorus DS only applied in-furrow at a rate of 5 lbs per acre.
4. APP + EnzUP Phosphorus DS applied in-furrow at a rate of 3 gallons per acre APP and 2 lbs per acre of EnzUP P DS. Rates were reduced for both products to apply same total P as treatments 2 and 3.
5. APP + EnzUP Potassium DS applied in-furrow at a rate of 5 gallons per acre APP and 5 lbs per acre of EnzUP K DS.
6. APP + EnzUP Zinc applied in-furrow at a rate of 5 gallons per acre APP and 1 quart per acre of EnzUP Zinc.
7. APP + EnzUP Seedflow Zinc applied in-furrow at a rate of 5 gallons per acre APP and 2 ounces per 80,000 seed unit of EnzUP Seedflow Zinc.

RESULTS:

TABLE 10. *Corn grain moisture and grain yield differences observed from applied treatments in 2023. West Lafayette, IN 2023.*

TREATMENT	GRAIN MOISTURE	GRAIN YIELD
	--- % ---	-- bu/ac --
Nontreated Control (C)	22.9 a	288.6 d*
Ammonium Polyphosphate (APP; 10-34-0)	23.7 a	298.2 bcd
EnzUP Phosphorus DS	23.7 a	297.4 bcd
APP + EnzUP Phosphorus DS	23.2 a	294.2 cd
APP + EnzUP Potassium DS	23.5 a	305.8 a
APP + EnzUP Zinc	23.4 a	311.8 a
APP + EnzUP Seedflow Zinc	22.8 a	299.8 bc
P-value	*0.30*	*0.02*

* Mean values that do not contain the same corresponding letter are determined statistically different ($P < 0.1$). Columns with mean values that do not contain any letters are determined as no statistical differences between treatments.

SUMMARY (TAKE-HOME POINTS):

- The inclusion of an in-furrow application of ammonium polyphosphate only (APP; 10-34-0) did not increase corn grain yield in comparison to the control (Table 10).
- Corn grain yield was significantly increased in comparison to the control with the inclusion of APP + EnzUP Potassium DS, APP + EnzUP Zinc, and APP + EnzUP Seedflow Zinc.

EVALUATION OF UREASE INHIBITOR ANVOL® WITH SURFACE-APPLIED UAN + ATS IN CORN (ACRE)

Daniel Quinn: Department of Agronomy, Purdue University

Rachel Stevens: Purdue Agronomy Center for Research and Education (ACRE)

Study Location: West Lafayette, IN

Soil Type: Drummer Fine-Silty (0–2% slope)

Planting Date: May 5, 2023 | **Harvest Date:** October 23, 2023

Corn Hybrid: Pioneer P1108Q | **Corn Seeding Rate:** 32,000 seeds/ac

Corn Nitrogen (N) Fertilizer Rate and Source: 0, 60, 120, 180 and 240 lbs N/ac, UAN (28-0-0). Nitrogen fertilizer was applied at the V1–V2 growth stage as a surface-band application between the corn rows and no starter N fertilizer was applied.

Previous Crop: Soybean | **Tillage:** Conventional

Study Replications: 4

RESEARCH TRIAL OVERVIEW:

This research trial was established at the Agronomy Center for Research and Education (ACRE) in Tippecanoe County, IN. This research trial examined corn grain yield differences in response to surface-banded UAN combined with ammonium thiosulfate (ATS) and/or the urease inhibitor ANVOL from Koch Agronomic Services. ATS was mixed with UAN at the time of application to equate a fertilizer analysis of 28-0-0-5S. Anvol was mixed with UAN at a rate of 0.75 quarts per ton. The experimental design of this trial was a randomized complete block design with four replications. All plots measured 15 feet wide (six 30-inch corn rows) x 40 feet long. The center two rows of each plot were harvested with a small-plot combine and adjusted to 15.5% moisture for yield analysis.

TABLE 11. *Mean corn grain yield in response to applied treatments and nitrogen fertilizer rates. West Lafayette, IN 2023.*

TREATMENT DESCRIPTION	NITROGEN FERTILIZER RATE	GRAIN MOISTURE	GRAIN YIELD
	---- lbs/ac ----	---- % ----	---- bu/ac ---
UAN + ATS	60	18.9 d*	232.8 e
	120	19.8 bcd	257.8 d
	180	20.9 abc	272.9 bc
	240	20.8 ab	281.6 ab
UAN + ATS + ANVOL	60	19.5 cd	242.0 e
	120	19.4 cd	266.8 cd
	180	20.8 a	280.2 ab
	240	20.7 ab	287.9 a
P-value		0.02	0.01
Untreated Check[†]	0	19.2	192.2

* Mean values that do not contain the same corresponding letter are determined statistically different ($P < 0.1$). Columns with mean values that do not contain any letters are determined as no statistical differences between treatments.
† Untreated check was not included in the analysis.

TABLE 12. *Mean corn grain yield differences in response to the inclusion of ANVOL across all applied nitrogen fertilizer rates. West Lafayette, IN 2023.*

TREATMENT DESCRIPTION	GRAIN MOISTURE	GRAIN YIELD
	---- % ----	---- bu/ac ---
UAN + ATS	20.2 a*	261.3 b
UAN + ATS + ANVOL	19.9 a	269.2 a
P-value	0.54	0.03

* Mean values that do not contain the same corresponding letter are determined statistically different ($P < 0.1$). Columns with mean values that do not contain any letters are determined as no statistical differences between treatments.

SUMMARY (TAKE-HOME POINTS):

- Corn grain yield was increased by 8 bushels per acre with the inclusion of ANVOL urease inhibitor when examined across N fertilizer rates applied (Table 12). Application of UAN (28-0-0) in this experiment was surface-banded following emergence in May and minimal to no rainfall was received after application, which suggests high potential for N volatilization.

EVALUATION OF NITRIFICATION INHIBITOR CENTURO® WITH COULTER-INJECTED UAN + ATS IN CORN (ACRE)

Daniel Quinn: Department of Agronomy, Purdue University

Rachel Stevens: Purdue Agronomy Center for Research and Education (ACRE)

Study Location: West Lafayette, IN

Soil Type: Drummer Fine-Silty (0–2% slope)

Planting Date: May 5, 2023 | **Harvest Date:** October 23, 2023

Corn Hybrid: Pioneer P1108Q | **Corn Seeding Rate:** 32,000 seeds/ac

Corn Nitrogen (N) Fertilizer Rate and Source: 0, 60, 120, 180 and 240 lbs N/ac, UAN (28-0-0). Nitrogen fertilizer was applied at the V1–V2 growth stage and coulter-injected between the rows. No starter fertilizer N was applied in this study.

Previous Crop: Soybean | **Tillage:** Conventional

Study Replications: 4

RESEARCH TRIAL OVERVIEW:

This research trial was established at the Agronomy Center for Research and Education (ACRE) in Tippecanoe County, IN. This research trial examined corn grain yield differences in response to coulter-injected UAN combined with ammonium thiosulfate (ATS) and/or the nitrification inhibitor Centuro from Koch Agronomic Services. ATS was mixed with UAN at the time of application to equate a fertilizer analysis of 28-0-0-5S. Centuro was mixed with UAN at a rate of 1.5 gallons per ton. The experimental design of this trial was a randomized complete block design with four replications. All plots measured 15 feet wide (six 30-inch corn rows) x 40 feet long. The center two rows of each plot were harvested with a small-plot combine and adjusted to 15.5% moisture for yield analysis.

TABLE 13. *Mean corn grain yield in response to applied treatments and nitrogen fertilizer rates. West Lafayette, IN 2023.*

TREATMENT DESCRIPTION	NITROGEN FERTILIZER RATE	GRAIN MOISTURE	GRAIN YIELD
	---- lbs/ac ----	---- % ----	---- bu/ac ---
UAN + ATS	60	19.1 b*	217.5 e
	120	18.9 b	252.7 d
	180	23.5 a	276.9 b
	240	22.7 ab	274.1 bc
UAN + ATS + CENTURO	60	19.3 b	216.9 e
	120	22.1 ab	257.8 cd
	180	22.7 ab	288.8 ab
	240	20.6 ab	300.9 a
P-value		0.09	0.01
Untreated Check†	0	18.5	172.6

* Mean values that do not contain the same corresponding letter are determined statistically different (*P* < 0.1). Columns with mean values that do not contain any letters are determined as no statistical differences between treatments.
† Untreated check was not included in the analysis.

TABLE 14. *Mean corn grain yield differences in response to the inclusion of CENTURO across all applied nitrogen fertilizer rates. West Lafayette, IN 2023.*

TREATMENT DESCRIPTION	GRAIN MOISTURE	GRAIN YIELD
	---- % ----	---- bu/ac ---
UAN + ATS	21.0	255.3 b*
UAN + ATS + CENTURO	21.1	266.1 a
P-value	0.64	0.05

* Mean values that do not contain the same corresponding letter are determined statistically different (*P* < 0.1). Columns with mean values that do not contain any letters are determined as no statistical differences between treatments.

SUMMARY (TAKE-HOME POINTS):

- Across all N fertilizer rates applied, the inclusion of CENTURO increased corn grain yield by 11 bushels per acre (Table 14).
- The greatest response to CENTURO was observed at the highest N fertilizer rates applied (Table 13). At an N fertilizer rate of 240 lbs N/ac, CENTURO improved corn grain yield by 26 bushels per acre.

CORN RESPONSE TO INSTINCT NXTGEN® ACROSS N FERTILIZER SOURCES, RATES, AND TIMINGS (ACRE)

Daniel Quinn: Department of Agronomy, Purdue University
Rachel Stevens: Purdue Agronomy Center for Research and Education (ACRE)

Study Location: West Lafayette, IN
Soil Type: Drummer Fine-Silty (0–2% slope)
Planting Date: May 19, 2023 | **Harvest Date:** October 23, 2023
Corn Hybrid: Pioneer P1108Q | **Corn Seeding Rate:** 32,000 seeds/ac
Corn Nitrogen (N) Fertilizer Rate and Source: 0, 160, and 180 lbs N/ac, UAN (28-0-0; Sidedress and Split Applications Only) and Urea (46-0-0; Preplant Applications Only). Split N applications included 50% N applied at planting and 50% N applied at the V5-6 growth stage. No starter fertilizer N was included in this study.
Previous Crop: Soybean | **Tillage:** Conventional
Study Replications: 4

RESEARCH TRIAL OVERVIEW:

This research trial was established at the Agronomy Center for Research and Education (ACRE) in Tippecanoe County, IN. This research trial examined corn grain yield differences in response to various N fertilizer sources, rates, and timings with and without nitrification inhibitor Instinct NXTGEN from Corteva Agriscience. The experimental design of this trial was a randomized complete block design with four replications. All plots measured 15 feet wide (six 30-inch corn rows) x 40 feet long. The center two rows of each plot were harvested with a small-plot combine and adjusted to 15.5% moisture for yield analysis.

TABLE 15. *Mean corn yield averages examining the interaction between total N fertilizer rate and Instinct NXTGEN. West Lafayette, IN 2023.*

NITROGEN FERTILIZER RATE	INSTINCT NXTGEN	GRAIN MOISTURE	GRAIN YIELD
---- lbs/ac ----		---- % ----	---- bu/ac ----
160	No	21.7	285.5*
160	Yes	21.3	285.6
180	No	21.6	285.5
180	Yes	21.7	290.9
P-value		0.20	0.48
Untreated Check (0 lbs N/ac)†	—	—	182.4

* Mean values that do not contain the same corresponding letter are determined statistically different ($P < 0.1$). Columns with mean values that do not contain any letters are determined as no statistical differences between treatments.
† Untreated check was not included in the analysis.

TABLE 16. *Mean corn yield averages examining the interaction between N fertilizer source and Instinct NXTGEN. West Lafayette, IN 2023.*

NITROGEN FERTILIZER SOURCE	INSTINCT NXTGEN	GRAIN MOISTURE	GRAIN YIELD
---- lbs/ac ----		---- % ----	---- bu/ac ----
28% UAN	No	21.8	283.9*
28% UAN	Yes	21.5	288.6
Urea (46-0-0)	No	21.4	288.5
Urea (46-0-0)	Yes	21.4	287.4
P-value		0.497	0.386
Untreated Check (0 lbs N/ac)[†]	—	—	182.4

* Mean values that do not contain the same corresponding letter are determined statistically different ($P < 0.1$). Columns with mean values that do not contain any letters are determined as no statistical differences between treatments.
† Untreated check was not included in the analysis.

TABLE 17. *Mean corn yield averages examining the interaction between N fertilizer timing and Instinct NXTGEN. West Lafayette, IN 2023.*

NITROGEN FERTILIZER TIMING	INSTINCT NXTGEN	GRAIN MOISTURE	GRAIN YIELD
---- lbs/ac ----		---- % ----	---- bu/ac ----
Preplant	No	21.4	288.5 a*
Preplant	Yes	21.4	287.4 ab
V2 Sidedress	No	22.1	288.7 a
V2 Sidedress	Yes	21.7	287.7 ab
Split Application	No	21.5	279.3 b
Split Application	Yes	21.4	289.5 a
P-value		0.669	0.078
Untreated Check (0 lbs N/ac)[†]	—	—	182.4

* Mean values that do not contain the same corresponding letter are determined statistically different ($P < 0.1$). Columns with mean values that do not contain any letters are determined as no statistical differences between treatments.
† Untreated check was not included in the analysis.

SUMMARY (TAKE-HOME POINTS):

- The inclusion of Instinct NXTGEN did not increase corn grain yield across both N fertilizer rates (160 and 180 lbs N/acre) and N fertilizer sources (UAN and urea) in this research trial (Tables 15 and 16).
- The inclusion of Instinct NXTGEN increased corn grain yield by 10 bushels per acre when applied as a split application (50% N applied at planting and 50% N applied at the V5–V6 growth stage; Table 17). Early-season dry conditions and higher rainfall amounts observed following the later (V5–V6) in-season sidedress application suggests this application method had the highest potential for N loss, thus likely driving the observed response.

COMPARISON OF DENT CORN AND POPCORN RESPONSES TO NITROGEN FERTILIZER RATES (ACRE)

Daniel Quinn: Department of Agronomy, Purdue University

Narciso Zapata: Department of Agronomy, Purdue University

Evan Bossung: Purdue Agronomy Center for Research and Education (ACRE)

Aaron Kult: Purdue Agronomy Center for Research and Education (ACRE)

Rachel Stevens: Purdue Agronomy Center for Research and Education (ACRE)

Study Location: West Lafayette, IN

Soil Type: Chalmers silty clay loam (0–2% slope)

Planting Date: May 12, 2023 | **Harvest Date:** November 10, 2023

Corn Hybrid: Pioneer P1108Q and Weaver W2021 | **Corn Seeding Rate:** 34,000 seeds/ac

Corn Nitrogen (N) Fertilizer Rate and Source: 0, 60, 120, 180, and 240 lbs N/ac, UAN (28-0-0). No starter fertilizer N was applied in this trial.

Previous Crop: Soybean | **Tillage:** Conventional

Study Replications: 4

RESEARCH TRIAL OVERVIEW:

This research trial was established at the Agronomy Center for Research and Education (ACRE) in Tippecanoe County, IN. This research trial examined both dent corn and popcorn grain yield differences in response to various N fertilizer rates. Nitrogen fertilizer was applied as 28% UAN at all locations and was coulter-injected as sidedress at the V_5 growth stage. Dent corn and popcorn were planted side by side in the same field and managed the exact same way. The experimental design of this trial was a randomized complete block design with four replications. All plots measured 30 feet wide (12, 30-inch corn rows) x 400 feet long. The center eight rows of each plot were harvested with a commercial combine containing a calibrated yield monitor.

TABLE 18. *Popcorn and dent corn grain yield and agronomic efficiency (AE) response to applied N fertilizer rates. West Lafayette, IN 2023.*

CORN TYPE	N FERTILIZER RATE	GRAIN YIELD	NAE[†]
		--- bu/ac ---	--- lbs/lb ---
Dent	0	146.2 d*	—
	60	228.5 c	76.8 a
	120	256.8 b	51.6 b
	180	268.1 ab	37.9 c
	240	272.5 a	29.5 d
		--- lbs/ac ---	-- lbs/b --
Popcorn	0	3808.9 c	—
	60	6293.4 b	41.4 c
	120	7400.1 a	29.9 d
	180	7302.3 a	19.4 e
	240	7430.3 a	15.1 e
P–value		*0.01*	*0.01*

* Mean values that do not contain the same corresponding letter are determined statistically different ($P < 0.1$).

[†] Nitrogen Agronomic Efficiency, NAE. Total lbs of grain harvested per lb of N fertilizer applied. Calculated as (yield − yield at 0 N)/N fertilizer rate applied). AE is analyzed across corn types to compare corn type differences, and mean values within the same corn type that do not contain a similar letter are statistically different ($P < 0.1$). Grain yield is separated by corn type due to discrepancies in units used for grain yield and discrepancies in total yield levels. Any mean yield values within each individual corn type that do not contain the same letter are statistically different ($P < 0.1$).

SUMMARY (TAKE-HOME POINTS):

- Agronomic optimum nitrogen rate (AONR) for popcorn was calculated as 132 lbs N/ac, whereas the AONR for dent corn was 152 lbs N/ac at this location (Table 18).
- When compared across N rates applied, dent corn can produce more lbs of grain per lb of N fertilizer applied than popcorn when grown in the same environment.

CORN RESPONSE TO ASYMBIOTIC N-FIXING BIOINOCULANT PRODUCTS (ACRE)

Daniel Quinn: Department of Agronomy, Purdue University
Narciso Zapata: Department of Agronomy, Purdue University
Roland Wilhelm: Department of Agronomy, Purdue University
Rachel Stevens: Purdue Agronomy Center for Research and Education (ACRE)

Study Location: West Lafayette, IN
Soil Type: Drummer silty clay loam (0–2% slope), Milford silty clay loam (0–2% slope)
Planting Date: May 05, 2023 | **Harvest Date:** October 12, 2023
Corn Hybrid: Pioneer P1108Q | **Corn Seeding Rate:** 32,000 seeds/ac
Corn Nitrogen (N) Fertilizer Rate and Source: 0, 60, 120, and 180 lbs N/ac, UAN (28-0-0) and Ammonium Thiosulfate (12-0-0-26).
Previous Crop: Soybean | **Tillage:** Conventional
Study Replications: 4

RESEARCH TRIAL OVERVIEW:

A field research trial was established at the Purdue Agronomy Center for Research and Education (ACRE) in Tippecanoe County, IN. The research trial examined corn yield response to different biological products application and N rates to analyze the impact of the interaction between both factors. The trial was designed as a randomized complete block design with 16 treatments and four replications. Plots measured 15 feet wide (six 30-inch corn rows) by 40 feet long, and the center two rows were harvested with a small-plot combine and adjusted to 15.5% moisture for yield analysis.

RESEARCH TREATMENTS:

BIOLOGICAL PRODUCT:
1. No biological
2. Envita foliar applied at V6 (Azotic NA)
3. Utrisha foliar applied at V6 (Corteva Agrisciences)
4. Proven40 in-furrow applied at planting (Pivot Bio)

NITROGEN FERTILIZER RATE:
1. 0 lb N/ac
2. 60 lb N/ac
3. 120 lb N/ac
4. 180 lb N/ac

RESULTS:

TABLE 19. *Mean plant population, grain moisture, and grain yield differences observed across biological products and nitrogen (N) fertilizer application rate in 2023. West Lafayette, IN.*

NITROGEN RATE	BIOLOGICAL PRODUCT	V5 PLANT POPULATION	GRAIN MOISTURE	GRAIN YIELD
-- lb/ac --		-- plants/ac --	-- % --	-- bu/ac --
0	None	31,610*	15.6*	141.8*
	Envita	31,281	16.3	135.7
	Utrisha	32,321	15.4	119.9
	Proven40	31,391	15.9	151.4
60	None	31,336	17.5	209.4
	Envita	31,828	17.2	200.2
	Utrisha	31,719	17.1	208.7
	Proven40	32,157	17.6	211.0
120	None	31,571	18.6	233.4
	Envita	31,774	17.8	233.7
	Utrisha	32,266	17.9	239.4
	Proven40	31,883	18.3	234.9
180	None	32,321	18.8	247.8
	Envita	31,992	19.4	251.4
	Utrisha	31,172	19.2	245.2
	Proven40	31,665	18.8	250.7
P-value		*0.17*	*0.86*	*0.70*

*Mean values that do not contain the same corresponding letter are determined statistically different ($P < 0.1$). Columns with mean values that do not contain any letters are determined as no statistical differences between treatments.

SUMMARY (TAKE-HOME POINTS):

- Corn plant stand was not responsive to N rate nor bioinoculant product across all treatments (Table 19) in West Lafayette, IN.
- Preliminary results suggest application of asymbiotic N-fixing bioinoculants do not improve corn yield or reduce synthetic N fertilizer requirements for corn at this specific location. However, analysis of additional locations is still ongoing and trials will be expanded to more locations in 2024 to help determine repeatability of results.

PINNEY PURDUE AGRICULTURAL CENTER (PPAC)

CORN RESPONSE TO VARIOUS IN-FURROW AND SEED-APPLIED NUTRIENT PRODUCTS (PPAC)

Daniel Quinn: Department of Agronomy, Purdue University
Alex Leman: Pinney Purdue Agricultural Center (PPAC)
Stephen Boyer: Pinney Purdue Agricultural Center (PPAC)

Study Location: Wanatah, IN
Soil Type: Sebewa Loam
Planting Date: May 19, 2023 | **Harvest Date:** October 23, 2023
Corn Hybrid: Pioneer P1108Q | **Corn Seeding Rate:** 32,000 seeds/ac
Corn Nitrogen (N) Fertilizer Rate and Source: 200 lbs N/ac, UAN (28-0-0). No additional starter fertilizer N was included in this study.
Previous Crop: Soybean | **Tillage:** Conventional
Study Replications: 4

RESEARCH TRIAL OVERVIEW:

This research trial was established at the Pinney Purdue Agricultural Center (PPAC) in Porter County, IN. This research trial examined corn grain yield differences to in-furrow and seed-applied nutritional products from Brandt. The experimental design of this trial was a randomized complete block design with four replications. All plots measured 30 feet wide (12, 30-inch corn rows) x 500+ feet long. The center six rows were harvested with a commercial combine containing a calibrated yield monitor and adjusted to 15.5% moisture for yield analysis.

RESEARCH TREATMENTS:

1. Control treatment (C) with no in-furrow starter fertilizer product applied.
2. Ammonium Polyphosphate (APP; 10-34-0) applied in-furrow at a rate of 5 gallons per acre.
3. EnzUP Phosphorus DS only applied in-furrow at a rate of 5 lbs per acre.

4. APP + EnzUP Phosphorus DS applied in-furrow at a rate of 3 gallons per acre APP and 2 lbs per acre of EnzUP P DS. Lower product rates were used to maintain the same total P as treatments 2 and 3.

5. APP + EnzUP Potassium DS applied in-furrow at a rate of 5 gallons per acre APP and 5 lbs per acre of EnzUP K DS.

6. APP + EnzUP Zinc applied in-furrow at a rate of 5 gallons per acre APP and 1 quart per acre of EnzUP Zinc.

7. APP + EnzUP Seedflow Zinc applied in-furrow at a rate of 5 gallons per acre APP and 2 ounces per 80,000 seed unit of EnzUP Seedflow Zinc.

RESULTS:

TABLE 20. *Corn grain moisture and grain yield differences observed from applied treatments in 2023. Wanatah, IN 2023.*

TREATMENT	GRAIN MOISTURE	GRAIN YIELD
	--- % ---	-- bu/ac --
Control (C)	22.5	272.4 bc[*]
Ammonium Polyphosphate (APP; 10-34-0)	22.6	268.1 c
EnzUP Phosphorus DS	22.7	273.1 abc
APP + EnzUP Phosphorus DS	22.5	272.6 bc
APP + EnzUP Potassium DS	22.7	271.6 bc
APP + EnzUP Zinc	22.7	278.7 a
APP + EnzUP Seedflow Zinc	22.7	276.3 ab
P-value	*0.463*	*0.013*

[*] Mean values that do not contain the same corresponding letter are determined statistically different ($P < 0.1$). Columns with mean values that do not contain any letters are determined as no statistical differences between treatments.

SUMMARY (TAKE-HOME POINTS):

• The inclusion of an in-furrow application of ammonium polyphosphate (APP; 10-34-0) only did not increase corn grain yield in comparison to the control (Table 20).

• Corn grain yield was significantly increased in comparison to the control with the inclusion of APP + EnzUP Zinc, and APP + EnzUP Seedflow Zinc.

CORN RESPONSE TO PLANTING DATE, HYBRID MATURITY, AND FUNGICIDE (PPAC)

Daniel Quinn: Department of Agronomy, Purdue University
Erick Oliva: Department of Agronomy, Purdue University
Stephen Boyer: Pinney Purdue Agricultural Center (PPAC)
Alex Leman: Pinney Purdue Agricultural Center (PPAC)
Steven Brand: Botany and Plant Pathology, Purdue University
Darcy Telenko: Botany and Plant Pathology, Purdue University

Study Location: Pinney Purdue Agricultural Center, Wanatah, IN
Soil Type: Sebewa loam
Planting Date: 1st Planting Date: May 18, 2023; 2nd Planting Date: June 1, 2023
Harvest Date: Oct. 27, 2023
Corn Hybrid(s): Pioneer P1108Q and Pioneer P9608Q
Corn Seeding Rate: 30,000 seeds/ac
Corn Nitrogen (N) Fertilizer Rate and Source: 212 lbs N per acre, UAN (28-0-0). 40 lbs N/ac was applied in a 2x2 starter at planting.
Corn Fungicide Product and Rate Used: Adastrio, 9 oz/ac, applied at the R1 growth stage.
Previous Crop: Corn | **Tillage:** Conventional
Study Replications: 4

RESEARCH TRIAL OVERVIEW:

A field research trial was established at the Pinney Purdue Agricultural Center (PPAC) in Porter County, IN. The research trial examined corn yield response to hybrid maturity, planting date, and foliar fungicide application applied at the R1 growth stage (silk emergence). The trial was designed as a randomized complete block design with four replications. Plots measured 30 feet wide (12, 30-inch corn rows) by 500+ feet long, and the center six rows were harvested with a commercial combine containing a calibrated yield monitor and adjusted to 15.5% moisture for yield analysis.

RESULTS:

TABLE 21. *Corn grain moisture and yield in response to hybrid, planting date, and fungicide. Wanatah, IN 2023.*

CORN HYBRID	PLANTING DATE	R1 FUNGICIDE APPLICATION	GRAIN MOISTURE	GRAIN YIELD
			%	bu/ac
Pioneer	18-May	Yes	18.63 ef*	229.62 c*
9608Q	18-May	No	18.35 f	227.36 dc
	1-June	Yes	19.56 d	222.69 d
	1-June	No	19.28 ed	221.64 d
Pioneer	18-May	Yes	22.07 c	258.6 a
1108Q	18-May	No	21.64 c	257.55 a
	1-June	Yes	25.16 a	248.98 b
	1-June	No	23.28 b	243.76 b
P-value			0.001	0.001

* Mean values that do not contain the same corresponding letter are determined statistically different ($P < 0.1$). Columns with mean values that do not contain any letters are determined as no statistical differences between treatments.

SUMMARY (TAKE-HOME POINTS):

- When comparing yield differences between the two hybrids selected and across both planting dates and fungicide applications, hybrid P1108Q (110-d) outyielded hybrid P9608Q (96-d) by an average of 27 bushels per acre (bu/ac) (Table 21).

- Fungicide application at the R1 growth stage (silk emergence) did not impact corn grain yield, regardless of hybrid maturity and/or planting date. Lack of fungicide response was likely due to minimal foliar disease observed at this location during the 2023 growing season. Foliar disease was found late in the growing season (R4–R5 growth stage) at low levels and was determined not to be yield limiting.

- The two corn hybrids examined did differ in grain yield response to planting date. The early maturing corn hybrid (P9608Q) had a yield increase with the early planting date. Corn planted on May 18 resulted in an average yield increase of 6 bu/ac for hybrid P9608Q as compared to a June 1 planting date. Similarly, the late-maturing corn hybrid (P1108Q) had a yield increase with the early planting date. Corn planted on May 18 resulted in an average yield increase of 12 bu/ac for hybrid P1108Q as compared to a June 1 planting date.

- For both corn hybrids examined, a later planting date resulted in a 1–2.5 percentage point grain moisture increase.

SHORT-STATURE CORN HYBRID RESPONSE TO ROW SPACING AND POPULATION (PPAC)

Daniel Quinn: Department of Agronomy, Purdue University

Erick Oliva: Department of Agronomy, Purdue University

Shaun Casteel: Department of Agronomy, Purdue University

Stephen Boyer: Pinney Purdue Agricultural Center (PPAC)

Alex Leman: Pinney Purdue Agricultural Center (PPAC)

Study Location: Pinney Purdue Agricultural Center, Wanatah, IN

Soil Type: Sebewa loam

Planting Date: May 23, 2023 | **Harvest Date:** November 9, 2023

Corn Hybrid: RT6203TVXZ, RV6205TVXZ, RW5419KTFZ | **Corn Seeding Rate:** 34,000; 42,000 and 50,000 seeds/ac

Corn Nitrogen (N) Fertilizer Rate and Source: 200 lbs N/ac, UAN (28-0-0). All plots received a broadcast N fertilizer application prior to planting. No starter fertilizer N was included in this trial.

Previous Crop: Soybean | **Tillage:** Conventional

Study Replications: 4

RESEARCH TRIAL OVERVIEW:

A field research trial was established at the Pinney Purdue Agricultural Center (PPAC) in La Porte County, IN. The research trial examined corn yield response to different short corn hybrids, row spacings, and seeding rates. The trial was designed as a split-plot, randomized complete block design with 18 treatments and four replications. Plots measured 10 feet wide (four 30-inch corn rows and six 20-inch corn rows) by 40 feet long, and the center two rows were harvested with a small-plot combine and adjusted to 15.5% moisture for yield analysis.

RESULTS:

TABLE 22. *Corn grain moisture and yield in response to row spacing, hybrid, and seeding rate. Wanatah, IN 2023.*

ROW SPACING	HYBRID	SEEDING RATE	GRAIN MOISTURE	GRAIN YIELD
		-- seeds/ac --	-- % --	-- bu/ac --
20-Inch Row	RT6203TVXZ	34,000	22.5 ab	257.3 b*
		42,000	22.7 a	271.3 a
		50,000	23.1 a	261.6 ab
	RV6205TVXZ	34,000	21.3 cd	252.1 b
		42,000	21.2 cde	249.5 b
		50,000	21.9 bc	279.8 a
	RW5419KTFZ	34,000	20.5 e	268.3 ab
		42,000	21.1 de	281.0 a
		50,000	20.9 de	278.7 a
30-Inch Row	RT6203TVXZ	34,000	22.3 a	247.9 a
		42,000	22.0 a	247.3 a
		50,000	22.0 a	248.9 a
	RV6205TVXZ	34,000	20.8 b	241.8 a
		42,000	21.2 b	245.4 a
		50,000	21.1 b	240.7 a
	RW5419KTFZ	34,000	20.6 b	262.9 a
		42,000	21.0 b	252.3 a
		50,000	21.1 b	253.2 a

* Mean values that do not contain the same corresponding letter are determined statistically different ($P < 0.1$). Columns with mean values that do not contain any letters are determined as no statistical differences between treatments. The data analyzed in this study was analyzed separately for each individual row spacing.

SUMMARY (TAKE-HOME POINTS):

- Short-stature corn hybrids illustrated a higher tolerance and optimum seeding rate in 20-in rows compared to 30-in rows, specifically for hybrids RT6203 and RV6205 (Table 22). Short-stature corn in 30-in rows showed no yield differences between seeding rates at Wanatah, IN.

- Preliminary data suggests short-stature corn hybrids can tolerate higher seeding rates in 20-in row spacing as compared to 30-in row spacings. In addition, hybrid yield differences may be dictated by ear height and compatibility with combine specifications. Therefore, short-stature hybrid selection that accounts for average ear height may be necessary to avoid any harvest challenges.

- When comparing yield differences between the three hybrids selected and across both row spacings and seeding rates, hybrid RW5419KTFZ outyielded hybrids RT6203TVXZ and RV6205TVXZ by an average of 10.5 and 14.5 bushels per acre (bu/ac), respectively.

DAVIS PURDUE AGRICULTURAL CENTER (DPAC)

CORN RESPONSE TO NITROGEN FERTILIZER APPLICATION TIMING FOLLOWING A RYE COVER CROP (DPAC)

Daniel Quinn: Department of Agronomy, Purdue University

Riley Seavers: Department of Agronomy, Purdue University

Shalamar Armstrong: Department of Agronomy, Purdue University

Jeff Boyer: Davis Purdue Agricultural Center (DPAC)

Study Location: Davis Purdue Agricultural Center, Farmland, IN

Soil Type: Pewamo silty clay loam (0–1% slope), Blount silt loam, ground moraine (0–2% slope)

Planting Date: May 23, 2023 | **Harvest Date:** November 20, 2023

Corn Hybrid: Pioneer P1108Q | **Corn Seeding Rate:** 30,000 seeds/ac

Corn Nitrogen (N) Fertilizer Rate and Source: Total N rate applied was 200 lbs N per acre, UAN (28-0-0). 40 lbs N/ac was applied as a 2x2 starter at planting across all N timing treatments.

Previous Crop: Soybean | **Tillage:** No-Till

Study Replications: 4

RESEARCH TRIAL OVERVIEW:

This research trial was established at the Davis Purdue Agricultural Center in Randolph County, IN. This research trial assessed corn yield differences when utilizing different in-season N fertilizer sidedress application timings following a fall-planted cereal rye cover crop and no rye cover crop. The rye cover crop was fall drill-seeded at a rate of 45 lbs/ac and was chemically terminated with glyphosate two weeks before corn planting. All plots received a 2x2 starter nitrogen application at planting of 40 lbs N/ac. Plots measured 30 feet wide (12, 30-inch corn rows) x 700+ feet long. The center eight rows were harvested with a commercial combine containing a calibrated yield monitor and were adjusted to 15.5% moisture for yield analysis.

RESEARCH TREATMENTS:

1. No cover crop + V5 sidedress N (coulter-inject, UAN 28-0-0)
2. No cover crop + V10 sidedress N (surface-banded, UAN 28-0-0)

3. No cover crop + V5 + V10 sidedress N (60% remaining N at V5 and 40% remaining N at V10)

4. Cereal rye cover crop + V5 sidedress N (coulter-inject, UAN 28-0-0)

5. Cereal rye cover crop + V10 sidedress N (surface-banded, UAN 28-0-0)

6. Cereal rye cover crop + V5 + V10 sidedress N (60% remaining N at V5 and 40% remaining N at V10)

RESULTS:

TABLE 23. *Average cereal rye cover crop aboveground biomass and nutrient uptake values. Rye cover crop biomass was sampled immediately prior to spring termination. Farmland, IN.*

BIOMASS	CARBON	NITROGEN	C:N*
-- lbs/ac --	-- lbs/ac --	-- lbs/ac --	
406.8	168.8	8.9	18:1

* Carbon (C) to nitrogen (N) ratio of aboveground biomass at termination.

TABLE 24. *Mean plant population, grain moisture, and grain yield differences observed across cereal rye cover crop presence and nitrogen (N) fertilizer application timing. Farmland, IN.*

COVER CROP	N TIMING	V5 PLANT POPULATION	GRAIN MOISTURE	GRAIN YIELD
		-- plants/ac --	-- % --	-- bu/ac --
No Cover	0N	28,268	21.1 a*	94.4 e
	V5	27,441	21.1 a	194.4 ab
	V10	28,216	20.9 a	172.4 c
	V5+V10	27,998	20.9 a	198.9 a
Rye Cover Crop	0N	28,218	21.2 a	75.1 f
	V5	28,649	21.1 a	199.4 a
	V10	27,830	20.9 a	159.1 d
	V5+V10	27,807	20.9 a	189.1 b
P-value		0.934	0.514	0.001

* Mean values that do not contain the same corresponding letter are determined statistically different (P < 0.1). Columns with mean values that do not contain any letters are determined as no statistical differences between treatments.

SUMMARY (TAKE-HOME POINTS):

- Rye cover crop biomass at termination averaged 407 lbs/ac, which contained 169 lbs/ac of carbon, 9 lbs/ac of N, and a C:N ratio of 18:1 (Table 23).
- Across all treatments examined, a rye cover crop reduced corn yield by 10 bu/ac.
- A significant interaction (P < 0.001) was observed between rye cover crop presence and N timing, which indicates optimum N timing did change with cover crop presence at this location. The V10 and V5+V10 N timings yielded less following a rye cover crop (Table 24).
- The V5 N application timing at DPAC had the highest yield value when following a rye cover crop (199 bu/ac) and resulted in comparable yield to treatments without a rye cover crop.
- Preliminary results suggest a full rate sidedress N application at the V5 growth stage is required to reach maximum yield potential when following a rye cover crop at this location.

CORN YIELD RESPONSE TO IN-SEASON NITROGEN (N) RATES ESTIMATED FROM SATELLITE IMAGERY (DPAC)

Daniel Quinn: Department of Agronomy, Purdue University

Ana Morales-Ona: Department of Agronomy, Purdue University

Jeff Boyer: Davis Purdue Agricultural Center (DPAC)

Study Location: Davis Purdue Agricultural Center, Farmland, IN

Soil Type: Blount silt loam, Pewamo silty clay loam, and Glynwood silt loam

Planting Date: May 19, 2023 | **Harvest Date:** November 9, 2023

Corn Hybrid: Pioneer P1108Q | **Corn Seeding Rate:** 30,000 seeds/ac

Farmer's Normal N Rate (FNR): 160 (w/o starter N) | **Starter N:** 40 | **Total N:** 200 lbs N per acre

Previous Crop: Corn | **Tillage:** Strip-till

RESEARCH TRIAL OVERVIEW:

This NRCS (Natural Resources Conservation Service) funded study examines the feasibility of using satellite imagery to determine maize N status and mid-season optimum N fertilizer rates. Five N fertilizer treatments were established and applied before planting based on the percentage of the farmer's N rate (FNR): (1) 40%, (2) 60%, (3) 80%, (4) 100%, and (5) 120% of the FNR. An additional treatment equivalent to the farmer's practice (starter and sidedress N only, no preplant N) was established too. Plots were 30 feet wide (12, 30-inch corn rows) by the length of the field. Treatments FP, 80, 100, and 120% FNR were replicated four times in a randomized complete block design. Treatments 40 and 60% FNR were replicated six times and placed adjacent to each side of the blocks with the FP, 80, 100, and 120% FNR treatments. Plots were further delineated into shorter sections, "subblocks" equal to the plot width by 200 ft long. Subblocks representing the range of all N rates were considered as a "block." At growth stage V12, variable-rate N fertilizer prescriptions were developed through identification of the agronomic optimum N fertilizer rate (AONR) of each block based on NDVI from satellite (PlanetScope multispectral images, 3-m resolution). Sidedress N was applied in the form of UAN (28%) in the areas corresponding to the treatments 40, 60, and 80% FNR. Field was harvested with a commercial combine and adjusted to 15.5% moisture for yield analysis.

RESULTS:

TABLE 25 *Nitrogen (pounds per acre) applications and mean grain yield (bushels per acre) per treatment. Farmland, IN 2023.*

TREATMENT	PREPLANT*	STARTER	SIDEDRESS	TOTAL	GRAIN YIELD
	----------- lbs N/ac -----------				---- bu/ac ----
Farmer's practice	0	40	161	201	136 c[†]
40% FNR sidedress 1 + sidedress 2	65	40	68	173	161 b
60% FNR sidedress 1 + sidedress 2	97	40	39	176	147 c
80% FNR sidedress 1 + sidedress 2	131	40	0	171	142 c
100% FNR sidedress	162	40	0	202	180 a
120% FNR sidedress	192	40	0	232	177 a
P–value					<0.001

* Preplant (UAN28 %): April 27; Starter (liquid mix): May 19; Sidedress (UAN 28%): July 24 ~V15.

[†] Mean values that do not contain the same corresponding letter are determined statistically different ($P < 0.1$). Columns with mean values that do not contain any letters are determined as no statistical differences between treatments.

SUMMARY (TAKE-HOME POINTS):

- Total N applied across the treatments ranged from 171 to 232 lbs N/ac, with 201 lbs N/ac being the farmer's normal total N rate applied (FNR; Table 25).
- For the treatments that received sidedress N application (40, 60, and 80% FNR + sidedress), total N applied ranged from 171 to 176 lbs N/ac. However, yield response to the 40% FNR + sidedress treatment was greater (161 bu/ac) compared to the other two sidedress treatments (147 and 142 bu/ac).
- Across all treatments examined, the 100% and 120% FNR resulted in the highest yields observed.

CORN RESPONSE TO PROVEN40® OS ACROSS MULTIPLE N FERTILIZER RATES (DPAC)

Daniel Quinn: Department of Agronomy, Purdue University

Jeff Boyer: Davis Purdue Agricultural Center (DPAC)

Study Location: Davis Purdue Agricultural Center, Farmland, IN

Soil Type: Pewamo silty clay loam (0–1% slope), Blount silt loam, ground moraine (0–2% slope)

Planting Date: May 23, 2023 | **Harvest Date:** October 24, 2023

Corn Hybrid: Pioneer P1108Q | **Corn Seeding Rate:** 30,000 seeds/ac

Corn Nitrogen (N) Fertilizer Rate and Source: Total N rates applied were 80, 160, and 240 lbs N per acre, UAN (28-0-0). 40 lbs N/ac was applied in a 2x2 starter at planting across all treatments.

Previous Crop: Soybean | **Tillage:** Strip-till

Study Replications: 4

RESEARCH TRIAL OVERVIEW:

This research trial was established at the Davis Purdue Agricultural Center in Randolph County, IN. This research trial assessed corn yield differences in response to the inclusion of Proven40 OS seed treatment across multiple N fertilizer rates. All plots received a 2x2 starter nitrogen application at planting of 40 lbs N/ac. Plots measured 30 feet wide (12, 30-inch corn rows) x 700+ feet long. The center eight rows were harvested with a commercial combine containing a calibrated yield monitor and were adjusted to 15.5% moisture for yield analysis.

TABLE 26. *Corn yield interaction response to seed-applied Pivot Bio and nitrogen fertilizer rate. Farmland, IN 2023.*

PROVEN40 OS	NITROGEN FERTILIZER RATE (LBS/AC)	GRAIN YIELD (BU/AC)
No	80	197.7 c*
Yes	80	195.9 c
No	160	219.3 b
Yes	160	236.2 a
No	240	226.7 b
Yes	240	235.1 a
P-value		0.001

* Mean values that do not contain the same corresponding letter are determined statistically different ($P < 0.1$). Columns with mean values that do not contain any letters are determined as no statistical differences between treatments.

SUMMARY (TAKE-HOME POINTS):

- The only response to Proven40 OS was observed at the 160 and 240 lbs N/ac N fertilizer rates and not the 80 lbs N/ac rate. Corn grain yield was increased by 17 and 9 bushels per acre at 160 lbs N/ac and 240 lbs N/ac, respectively with the inclusion of Proven40 OS in comparison to no biological seed treatment applied.
- Preliminary results and data collection provide an unclear reasoning as to why a response was observed at the higher N rates applied and not the 80 lbs N/ac rate.

NORTHEAST PURDUE AGRICULTURAL CENTER (NEPAC)

CORN RESPONSE TO INPUT-INTENSIVE MANAGEMENT (NEPAC)

Daniel Quinn: Department of Agronomy, Purdue University
Malena Bartaburu: Department of Agronomy, Purdue University
Darcy Telenko: Botany and Plant Pathology, Purdue University
Chris Lake: Northeast Purdue Agriculture Center (NEPAC)

Study Location: Northeast Purdue Agricultural Center, Columbia City, IN
Soil Type: Mermill loam, Haskins loam, Morley clay loam (6–12% slope), Coesse silty clay loam, Glynwood loam
Planting Date: May 25, 2023 | **Harvest Date:** Nov. 10, 2023
Corn Hybrid(s): Pioneer P1108Q | **Corn Seeding Rate:** 30,000 and 36,000 seeds/ac
Corn Nitrogen (N) Fertilizer Rate and Source: Total N rate across treatments was 200 lbs N per acre, UAN (28-0-0). 40 lbs N/ac was applied in a 2x2 starter at planting.
Previous Crop: Soybean | **Tillage:** No-till
Study Replications: 3

RESEARCH TRIAL OVERVIEW:

A field research trial was established at the Northeast Purdue Agricultural Center (NEPAC) in Whitley County, IN. The research trial examined corn yield response to different management practices and inputs and was utilized to analyze the impact of each one applied individually and in combination. The trial was designed as a randomized complete block design with eight treatments and five replications. Plots measured 30 feet wide (12, 30-inch corn rows) x 400+ feet long. The center eight rows were harvested with a commercial combine containing a calibrated yield monitor and were adjusted to 15.5% moisture for yield analysis.

TREATMENTS:

1. Control treatment (C) based on Purdue University seed rate (30K seeds/ac) and nitrogen (N) fertilizer rate recommendations
2. C + surface banded starter (2x2) fungicide (flutriafol, Xyway LFR, 15 oz/ac)

3. C + 20% increase in corn seeding rate (36K seeds/ac)

4. C + sulfur fertilizer (5.2 gallons/ac as ammonium thiosulfate (ATS) at V_5 sidedress)

5. C + foliar micronutrients (zinc, manganese, and boron applied at the V6 growth stage)

6. C + late-season N application (starter N (2x2) + V_5 sidedress N (60% remaining N rate) + V_{10-12} growth stage sidedress N surface-banded with drop tubes on a sprayer (40% remaining N rate), total N rate remained the same as other treatments)

7. C + foliar fungicide applied at the R1 growth stage (prothioconazole, trifloxystrobin, fluopyram, Delaro Complete, 10 oz/ac)

8. Intensive treatment: All additional inputs (except the Smart KB treatment) and management practices applied together

9. C + foliar fungicide applied at R1 + foliar fertilizer (SmartKB, 2qts/ac)

RESULTS:

TABLE 27. *Corn grain yield, grain moisture, treatment cost, and net profit differences observed from applied treatments in 2023. Columbia City, IN 2023.*

TREATMENT	GRAIN MOISTURE	GRAIN YIELD	TREATMENT COSTS§	NET PROFIT§
	--- % ---	-- bu/ac --	-- $/ac --	-- $/ac --
Control (C)	24.3	227.4	253.4	885.7
C + 2x2 Fungicide	NS*	+11.8	274.3	+38.5
C + Increased Seed	-1.1†	NS	272.7	NS
C + Sulfur	NS	+10.8	270.7	+37.1
C + Foliar Micro	NS	+18.4	274.2	+71.5
C + V_{10-12} SD N	NS	+14.5	261.1	+64.9
C + R1 Fungicide	+1.3	+25.9	277.9	+105.4
Intensive	+1.4	+22.1	356.1	NS
C + R1 Fungicide + SmartKB	+1.5	+31.0	288.9	+119.9
P-value	0.01	0.01	—	0.04

* NS, nonsignificant when compared to the control treatment using single degree of freedom contrasts at α = 0.1.

† Mean values that are present within columns indicate statistically significant increase or decrease in the examined variable from the control (α = 0.1).

§ Treatment costs were calculated as the combined cost of corn seed, fertilizer cost, chemical input cost, and application cost. Prices were calculated as an average from various local retailers. Net profit was calculated based on average harvest corn grain cash price ($5.01) + average grain yield – treatment costs.

SUMMARY (TAKE-HOME POINTS):

• The intensive management treatment outyielded the control by 22 bu/ac at this location (Table 27). The individual inputs with yield responses included Xyway LFR (+12 bu/ac), sulfur (+11 bu/ac), foliar micro (+18 bu/ac), late-season N (+14 bu/ac), R1 fungicide (+24 bu/ac), and R1 fungicide + SmartKB (+29 bu/ac).

• Due to the high application cost, the intensive treatment did not increase net profit compared to the control at this location.

- No yield response was observed from the 20% increase in seeding rate at this location.
- Preliminary results suggest the R1 fungicide had the greatest potential to increase corn grain yield at this location in 2023. The observed response was likely due to observed foliar disease presence and observed foliar disease suppression by the fungicide application at this location in 2023.

SOUTHEAST PURDUE AGRICULTURAL CENTER (SEPAC)

CORN RESPONSE TO INPUT-INTENSIVE MANAGEMENT (SEPAC)

Daniel Quinn: Department of Agronomy, Purdue University
Malena Bartaburu: Department of Agronomy, Purdue University
Darcy Telenko: Botany and Plant Pathology, Purdue University
Joel Wahlman: Southeast Purdue Agricultural Center (SEPAC)

Study Location: Southeast Purdue Agricultural Center, Butlerville, IN
Soil Type: Cobbsfork silt loam
Planting Date: May 23, 2023 | **Harvest Date:** November 3, 2023
Corn Hybrid(s): Dekalb DKC67-44
Corn Nitrogen (N) Fertilizer Rate and Source: Total N rate across treatments was 210 lbs N per acre, UAN (28-0-0). 40 lbs N/ac was applied in a 2x2 starter at planting.
Previous Crop: Soybean | **Tillage:** No-till
Study Replications: 5

RESEARCH TRIAL OVERVIEW:

A field research trial was established at the Southeast Purdue Agricultural Center (SEPAC) in Jennings County, IN. The research trial examined corn yield response to different management practices and inputs and was utilized to analyze the impact of each one applied individually and in combination. The trial was designed as a randomized complete block design with nine treatments and five replications. Plots measured 30 feet wide (12, 30-inch corn rows) x 500+ feet long. The center six rows were harvested with a commercial combine containing a calibrated yield monitor and were adjusted to 15.5% moisture for yield analysis.

TREATMENTS:

1. Control treatment (C) based on Purdue University seed rate (30K seeds/ac) and nitrogen (N) fertilizer rate recommendations
2. C + surface banded starter (2x2x2) fungicide (flutriafol, Xyway LFR, 15 oz/ac)

3. C + 20% increase in corn seeding rate (36K seeds/ac)

4. C + sulfur fertilizer (5.2 gallons/ac as ammonium thiosulfate (ATS) at V_5 sidedress)

5. C + foliar micronutrients (zinc, manganese, and boron applied at the V6 growth stage)

6. C + late-season N application (starter N (2x2x2) + V_5 sidedress N (60% remaining N rate) + V10-12 growth stage sidedress N surface-banded with drop tubes on a sprayer (40% remaining N rate), total N rate remained the same as other treatments)

7. C + foliar fungicide applied at the R1 growth stage (prothioconazole, trifloxystrobin, fluopyram, Delaro Complete, 10 oz/ac)

8. Intensive treatment: All additional inputs (except the Smart KB treatment) and management practices applied together

9. C + foliar fungicide applied at R1 + foliar fertilizer (SmartKB, 2qts/ac)

RESULTS:

TABLE 28. *Corn grain yield, grain moisture, treatment cost, and net profit differences observed from applied treatments in 2023. Butlerville, IN 2023.*

TREATMENT	GRAIN MOISTURE	GRAIN YIELD	TREATMENT COSTS§	NET PROFITS§
	--- % ---	-- bu/ac --	-- \$/ac --	-- \$/ac --
Control	21.3 de†	250.8 de	253.4	1002.9
C + 2x2x2 Fungicide	NS*	NS	274.3	NS
C + Increased Seed	NS	+7.8	272.7	NS
C + Sulfur	NS	NS	270.7	NS
C + Foliar Micro	NS	NS	274.2	NS
C + V10-12 SD N	NS	NS	261.1	NS
C + R1 Fungicide	+0.5	+7.1	277.9	NS
Intensive	+0.4	+7.0	356.1	-67.4
C + R1 Fungicide + SmartKB	0.4	+12.6	288.9	+27.8
P-value	0.01	0.01	—	0.05

* NS, nonsignificant when compared to the control treatment using single degree of freedom contrasts at α = 0.1.

† Mean values that are present within columns indicate statistically significant increase or decrease in the examined variable from the control (α = 0.1).

§ Treatment costs were calculated as the combined cost of corn seed, fertilizer cost, chemical input cost, and application cost. Prices were calculated as an average from various local retailers. Net profit was calculated based on average harvest corn grain cash price (\$5.01) + average grain yield – treatment costs.

SUMMARY (TAKE-HOME POINTS):

- The intensive management treatment outyielded the control by 7 bu/ac at this location (Table 28). The individual inputs with yield responses included a 20% increase in seeding rate (+8 bu/ac), R1 fungicide (+7 bu/ac), and R1 fungicide + SmartKB (+13 bu/ac).

- No yield response was observed from the foliar micro, sulfur, and late-season N application in this study.

- Despite observed yield responses, no input significantly increased net profit in comparison to the control at this location in 2023.

CORN RESPONSE TO NITROGEN FERTILIZER APPLICATION TIMING FOLLOWING A RYE COVER CROP (SEPAC)

Daniel Quinn: Department of Agronomy, Purdue University

Riley Seavers: Department of Agronomy, Purdue University

Shalamar Armstrong: Department of Agronomy, Purdue University

Joel Wahlman: Southeast Purdue Agricultural Center (SEPAC)

Study Location: Southeast Purdue Agricultural Center, Butlerville, IN

Soil Type: Cobbsfork, Avonburg, Nabb—silt loams

Planting Date: May 15, 2023 | **Harvest Date:** October 17, 2023

Corn Hybrid: Pioneer P0953AM | **Corn Seeding Rate:** 30,000 seeds/ac

Corn Nitrogen (N) Fertilizer Rate and Source: Total N rate across N timing treatments was 200 lbs N per acre, UAN (28-0-0). 40 lbs N/ac was applied in a 2x2 starter at planting.

Previous Crop: Soybean | **Tillage:** No-Till

Study Replications: 3

RESEARCH TRIAL OVERVIEW:

This research trial was established at the Southeast Purdue Agricultural Center in Jennings County, IN. This research trial assessed corn yield differences when utilizing different in-season N fertilizer sidedress application timings following a fall-planted cereal rye cover crop and no rye cover crop. The rye cover crop was fall drill-seeded at a rate of 45 lbs/ac and was chemically terminated with glyphosate two weeks before corn planting. All plots received a 2x2 starter nitrogen application at planting of 40 lbs N/ac. Plots measured 30 feet wide (12, 30-inch corn rows) x 700+ feet long. The center six rows were harvested with a commercial combine containing a calibrated yield monitor and were adjusted to 15.5% moisture for yield analysis.

RESEARCH TREATMENTS:

1. No cover crop + V_5 sidedress N (coulter-inject, UAN 28-0-0)
2. No cover crop + V_{10} sidedress N (surface-banded, UAN 28-0-0)
3. No cover crop + V_5 + V_{10} sidedress N (60% remaining N at V_5 and 40% remaining N at V_{10})
4. Cereal rye cover crop + V_5 sidedress N (coulter-inject, UAN 28-0-0)
5. Cereal rye cover crop + V_{10} sidedress N (surface-banded, UAN 28-0-0)
6. Cereal rye cover crop + V_5 + V_{10} sidedress N (60% remaining N at V_5 and 40% remaining N at V_{10})

RESULTS:

TABLE 29. *Average cereal rye cover crop aboveground biomass and nutrient uptake values. Rye cover crop biomass was sampled immediately prior to spring termination. Butlerville, IN 2023.*

BIOMASS	CARBON	NITROGEN	C:N*
-- lbs/ac --	-- lbs/ac --	-- lbs/ac --	
1130.4	462.9	19.8	24:1

* Carbon (C) to nitrogen (N) ratio of aboveground biomass at termination.

TABLE 30. *Mean plant population, grain moisture, and grain yield differences observed across cereal rye cover crop presence and nitrogen (N) fertilizer application timing. Butlerville, IN 2023.*

COVER CROP	N TIMING	V5 PLANT POPULATION	GRAIN MOISTURE	GRAIN YIELD
		-- plants/ac --	-- % --	-- bu/ac --
No Cover	V5	26,307	20.7	262.2 a*
	V10	26,813	20.4	250.5 cd
	V5+V10	26,188	20.7	265.2 a
Rye Cover Crop	V5	26,611	21.1	261.2 ab
	V10	26,072	20.4	245.1 d
	V5+V10	26,190	20.8	256.1 bc
P-value		0.54	0.39	0.01

* Mean values that do not contain the same corresponding letter are determined statistically different ($P < 0.1$). Columns with mean values that do not contain any letters are determined as no statistical differences between treatments.

SUMMARY (TAKE-HOME POINTS):

- Rye cover crop biomass at termination averaged 1130 lbs/ac, which contained 462 lbs/ac of carbon, 19 lbs/ac of N, and a C:N ratio of 24:1 (Table 29).
- Across all N timing treatments examined, a rye cover crop reduced corn yield by an average of 5 bu/ac.
- When examining N fertilizer application timing differences, when not following a rye cover crop, either the V5 or V5+V10 N application timings produced the highest yields (Table 30). However, when following a rye cover crop, the V5-only sidedress was the only N timing to produce comparable yields to corn not following a rye cover crop. In addition, a V10-only sidedress N application reduced corn yield in comparison to sidedress applied at the V5 growth stage regardless of cover crop presence.
- Preliminary results suggest a sidedress N application at the V5 growth stage is required to reach maximum yield potential when following a rye cover crop at this location.

CORN EMERGENCE AND YIELD RESPONSE TO CLOSING WHEEL TYPE IN A RYE COVER CROP SYSTEM (SEPAC)

Daniel Quinn: Department of Agronomy, Purdue University
Riley Seavers: Department of Agronomy, Purdue University
Joel Wahlman: Southeast Purdue Agricultural Center (SEPAC)

Study Location: Southeast Purdue Agricultural Center, Butlerville, IN
Soil Type: Cobbsfork silt loam.
Planting Date: May 17, 2023 | **Harvest Date:** October 17, 2023
Corn Hybrid: Pioneer 1136AM | **Corn Seeding Rate:** 30,000 seeds/ac
Corn Nitrogen (N) Fertilizer Rate and Source: Total N rate applied across treatments was 200 lbs N per acre, UAN (28-0-0). 40 lbs N/ac was applied in a 2x2 starter at planting and remaining N was sidedressed at the V5 growth stage.
Rye Cover Crop: VNS Cereal Rye, fall drill-seeded at 45 lbs per acre, chemically terminated 3 weeks prior to corn planting.
Previous Crop: Soybean | **Tillage:** No-till
Study Replications: 5

RESEARCH TRIAL OVERVIEW:

A field research trial was established at the Southeast Purdue Agricultural Center (SEPAC) in Jennings County, IN. The research trial examined corn emergence and yield response to planter closing wheel type in a no-till system with and without a rye cover crop. The trial was designed as a randomized complete block design with four replications. Plots measured 15 feet wide (six 30-inch corn rows) by 200+ feet long, and the center six rows were harvested with a commercial combine and adjusted to 15.5% moisture for yield analysis. All treatments received a 2x2 starter application of N fertilizer at planting totaling 40 lbs N/ac.

RESULTS:

TABLE 31. *Average cereal rye cover crop aboveground biomass and nutrient uptake values. Rye cover crop biomass was sampled immediately prior to spring termination. Butlerville, IN 2023.*

BIOMASS	CARBON	NITROGEN	C:N*
-- lbs/ac --	-- lbs/ac --	-- lbs/ac --	
1465.4	598.5	29.8	20:1

* Carbon (C) to nitrogen (N) ratio of aboveground biomass at termination.

TABLE 32. *Corn emergence, final stand, grain moisture, and yield in response to cover crop presence and closing wheel type. Butlerville, IN 2023.*

COVER CROP PRESENCE	CLOSING WHEEL TYPE*	%EMERGENCE (11 DAP*)	GRAIN MOISTURE	GRAIN YIELD
		%	%	bu/ac
No Cover Crop	Standard Rubber	92.3 b†	20.9 b	288.9 ab
	CA Cruiser Extreme	93.0 b	21.0 b	290.0 a
	MT Cupped Razor	98.7 a	21.0 b	289.7 a
Cereal Rye Cover	Standard Rubber	85.2 c	21.6 a	281.8 d
	CA Cruiser Extreme	97.4 a	21.5 a	286.7 bc
	MT Cupped Razor	91.1 b	21.4 a	285.3 c
P–value		0.01	0.01	0.04

* CA: Copperhead Ag, MT: Martin-Till, DAP: days after corn planting

† Mean values that do not contain the same corresponding letter are determined statistically different ($P < 0.1$). Columns with mean values that do not contain any letters are determined as no statistical differences between treatments.

SUMMARY (TAKE-HOME POINTS):

- The use of either the Cruiser Extreme or the Cupped Razor closing wheels improved corn emergence when following a RCC in comparison to the standard rubber closing wheel at 11 days after planting (DAP; Table 32).
- The inclusion of a rye cover crop increased corn grain harvest moisture by 0.6% across examined closing wheel treatments.
- The use of either the Cruiser Extreme or the Cupped Razor closing wheels improved corn yield when following a rye cover crop in comparison to the standard rubber closing wheel. Results suggest after-market closing wheels designed for high residue/cover crop systems (cast iron, spiked, or curve-toothed) can improve corn emergence and yield following a rye cover crop.

THROCKMORTON PURDUE AGRICULTURAL CENTER (TPAC)

CORN EMERGENCE AND YIELD RESPONSE TO CLOSING WHEEL TYPE IN A RYE COVER CROP SYSTEM (TPAC)

Daniel Quinn: Department of Agronomy, Purdue University
Riley Seavers: Department of Agronomy, Purdue University
Jay Young: Throckmorton Purdue Agricultural Center
Pete Illingsworth: Throckmorton Purdue Agricultural Center

Study Location: Throckmorton Purdue Agricultural Center, Lafayette, IN
Soil Type: Drummer silt loam, Throckmorton silt loam, and Toronto-Millbrook complex
Planting Date: May 18, 2023 | **Harvest Date:** November 14, 2023
Corn Hybrid: Pioneer 1108Q | **Corn Seeding Rate:** 30,000 seeds/ac
Corn Nitrogen (N) Fertilizer Rate and Source: Total N rate applied across treatments was 200 lbs N per acre, UAN (28-0-0). 40 lbs N/ac was applied in a 2x2 starter at planting and remaining N was sidedressed at the V5 growth stage.
Rye Cover Crop: VNS Cereal Rye, fall drill-seeded at 45 lbs per acre, chemically terminated 3 weeks prior to corn planting.
Previous Crop: Soybean | **Tillage:** No-till
Study Replications: 4

RESEARCH TRIAL OVERVIEW:

A field research trial was established at the Throckmorton Purdue Agricultural Center (TPAC) in Tippecanoe County, IN. The research trial examined corn emergence and yield response to planter closing wheel type in a no-till system with and without a rye cover crop. The trial was designed as a randomized complete block design with four replications. Plots measured 30 feet wide (12, 30-inch corn rows) by 700+ feet long, and the center six rows were harvested with a commercial combine and adjusted to 15.5% moisture for yield analysis. All treatments received a 2x2 starter application of N fertilizer at planting totaling 40 lbs N/ac.

RESULTS:

TABLE 33. *Average cereal rye cover crop aboveground biomass and nutrient uptake values. Rye cover crop biomass was sampled immediately prior to spring termination. Lafayette, IN 2023.*

BIOMASS	CARBON	NITROGEN	C:N*
-- lbs/ac --	-- lbs/ac --	-- lbs/ac --	
2503.1	1020.1	55.3	19:1

* Carbon (C) to nitrogen (N) ratio of aboveground biomass at termination.

TABLE 34. *Corn emergence, final stand, grain moisture, and yield in response to cover crop presence and closing wheel type. Lafayette, IN 2023.*

COVER CROP PRESENCE	CLOSING WHEEL TYPE*	%EMERGENCE (11 DAP*)	GRAIN MOISTURE	GRAIN YIELD
		%	%	bu/ac
No Cover Crop	Standard Rubber	103.3 a†	17.7 c	252.4 ab
	CA Cruiser Extreme	101.1 a	17.7 c	254.3 a
	MT Cupped Razor	97.9 a	17.9 b	255.1 a
Cereal Rye Cover	Standard Rubber	93.5 b	18.1 ab	244.9 b
	CA Cruiser Extreme	108.4 a	18.1 ab	252.5 ab
	MT Cupped Razor	103.9 ab	18.3 a	251.4 ab
P-value		0.03	0.04	0.04

* CA: Copperhead Ag, MT: Martin-Till, DAP: days after corn planting

† Mean values that do not contain the same corresponding letter are determined statistically different ($P < 0.1$). Columns with mean values that do not contain any letters are determined as no statistical differences between treatments.

SUMMARY (TAKE-HOME POINTS):

- The use of the Cruiser Extreme closing wheels improved corn emergence when following a RCC in comparison to the standard rubber closing wheel at 11 days after planting (DAP; Table 34).
- The inclusion of a rye cover crop increased corn grain harvest moisture by 0.5% across examined closing wheel treatments.
- The use of either the Standard Rubber, Cruiser Extreme, or the Cupped Razor closing wheels produced corn yields comparable to those without a rye cover crop. Results suggest after-market closing wheels designed for high residue/cover crop systems (cast iron, spiked, or curve-toothed) can improve corn emergence following a rye cover crop.

EVALUATION OF CORN EMERGENCE AND YIELD RESPONSE TO SEED DEPTH, HYBRID TYPE, AND ACTIVE DOWNFORCE (TPAC)

Daniel Quinn: Department of Agronomy, Purdue University

Pete Illingsworth: Throckmorton Purdue Agricultural Center

Tom Bechman: Indiana Prairie Farmer

Study Location: Throckmorton Purdue Agricultural Center, Lafayette, IN

Soil Type: Throckmorton silt loam (1–3% slope), Toronto-Millbrook complex (0–2% slope)

Planting Date: May 19, 2023 | **Harvest Date:** November 7, 2023

Corn Hybrid: Becks 6064AM and Becks 5909AM

Corn Seeding Rate: 30,000 seeds/ac

Corn Nitrogen (N) Fertilizer Rate and Source: Total N fertilizer rate across treatments was 210 lbs N per acre, UAN (28-0-0). 40 lbs N/ac was applied in a 2x2 starter at planting.

Previous Crop: Soybean | **Tillage:** Conventional

Study Replications: 3

RESEARCH TRIAL OVERVIEW:

A field research trial was established at the Throckmorton Purdue Agricultural Center (TPAC) in Tippecanoe County, IN. The research trial examined corn seedling emergence timing and yield differences between two different seeding depths and two different hybrids. The trial was designed as a randomized complete block design with three replications. Plots measured 30 feet wide (12, 30-inch corn rows) by 400+ feet long, and the center six rows were harvested with a commercial combine and adjusted to 15.5% moisture for yield analysis.

RESULTS:

TABLE 35. *Influence of hybrid type, depth, and downforce settings on corn plant population, moisture, and grain yield. Lafayette, IN 2023.*

TREATMENT	PLANT POPULATION	MOISTURE	GRAIN YIELD
	--- plants/acre ---	--- % ---	--- bu/ac ---
Hybrid 1 (Becks 5909AM)	29,208 a*	16.1 b	234.2 b
Hybrid 2 (Becks 6064AM)	28,792 a	16.9 a	249.8 a
2-inch depth	29,542 a	16.4 a	243.2 a
3-inch depth	28,458 b	16.4 a	240.8 a
100 lbs	28,583 b	16.5 a	239.9 b
Active	29,417 a	16.4 a	248.1 a

* Mean values that do not contain the same corresponding letter are determined statistically different ($P < 0.1$). Columns with mean values that do not contain any letters are determined as no statistical differences between treatments.

SUMMARY (TAKE-HOME POINTS):

- Hybrid 2 (Becks 6064AM) resulted in a higher yield (+15 bu/ac) when compared to hybrid 1 (Becks 5909AM; Table 35).
- No statistical yield differences were observed between the two seed depths examined. In addition, no interactions were observed between the different hybrids and seed depth and downforce settings examined.
- Active downforce increased corn yield by 9 bu/ac in comparison to static downforce set at 100 lbs. In addition, final plant stand was improved with the use of active downforce as compared to the static downforce setting. Furthermore, emergence timing and emergence uniformity was improved with active down force (data not shown), which may have contributed to the observed yield response.

ON-FARM RESEARCH TRIAL RESULTS

CORN YIELD RESPONSE TO IN-SEASON NITROGEN (N) RATES ESTIMATED FROM SATELLITE IMAGERY (WHITE CO.)

Daniel Quinn: Department of Agronomy, Purdue University

Ana Morales-Ona: Department of Agronomy, Purdue University

Study Location: White County, Indiana | **Field ID:** Carter Home

Soil Type: Alvin fine sandy loam, Foresman silt loam, Pella silty clay loam, Rensselaer clay loam, and Darroch silt loam

Planting Date: May 10, 2023 | **Harvest Date:** October 13, 2023

Corn Hybrid: DKC63-37 | **Corn Seeding Rate:** 35,500 seeds/ac

Farmer's normal N rate (FNR): 180 (w/o starter N) | **Extra N** (herbicide): 34 | **Total N:** 214 lbs N per acre

Previous Crop: Corn | **Tillage:** Conventional

RESEARCH TRIAL OVERVIEW:

This NRCS (Natural Resources Conservation Service) funded study examines the feasibility of using satellite imagery to determine maize N status and mid-season optimum N fertilizer rates. Five N fertilizer treatments were established and applied after planting (V2 growth stage) based on the percentage of the farmer's N rate (FNR): (1) 40%, (2) 60%, (3) 80%, (4) 100%, and (5) 120% of the FNR. Plots were 60 feet wide (24, 30-inch corn rows) by the length of the field. Treatments 80, 100, and 120% FNR were replicated two times in a randomized complete block design. Treatments 40 and 60% FNR were replicated three times and placed adjacent to the side of the blocks with the 80, 100, and 120% FNR treatments. A reduced number of replications were established for the higher N rate treatments (80, 100, and 120% FNR) with the aim to reduce overall total N rate applied. Plots were further delineated into shorter sections, "subblocks" equal to the plot width by 300 ft long. Subblocks representing the range of all N rates were considered as a "block." At growth stage V7–V8, variable-rate N fertilizer prescriptions were developed through identification of the agronomic optimum N fertilizer rate (AONR) of each block based on NDVI from satellite (PlanetScope multispectral images, 3-m resolution). A second sidedress N was applied in the form of UAN (28%) in the areas corresponding to the treatments of 40, 60, and 80% FNR. The field was harvested with a commercial combine and adjusted to 15.5% moisture for yield analysis.

RESULTS:

TABLE 36. *Nitrogen (pounds per acre) applications and mean grain yield (bushels per acre) per treatment. White County, IN 2023.*

TREATMENT	EXTRA N*	SIDEDRESS 1	SIDEDRESS 2	TOTAL	GRAIN YIELD
		----------- lbs N/ac -----------			---- bu/ac ----
40% FNR sidedress 1 + sidedress 2	34	72	72	178	193 ab[†]
60% FNR sidedress 1 + sidedress 2	34	108	36	178	188 ab
80% FNR sidedress 1 + sidedress 2	34	144	0	178	200 a
100% FNR sidedress	34	179	0	213	190 ab
120% FNR sidedress	34	215	0	249	181 b
P-value					*0.245*

* Extra N (UAN28 %) applied with herbicide; 1st sidedress (anhydrous NH3): May 18 ~V1; 2nd sidedress (UAN 28%): June 17 ~V7
[†] Mean values that do not contain the same corresponding letter are determined statistically different ($P < 0.1$).

SUMMARY (TAKE-HOME POINTS):

- Total N applied across the treatments ranged from 178 to 249 lbs N/ac, with 213 lbs N/ac being the farmer's normal total N rate applied (Table 36).
- For the treatments that received sidedress N application (40, 60, and 80% FNR + sidedress), total N applied was 178 lbs N/ac, and yield response was not statistically different.

CORN YIELD RESPONSE TO IN-SEASON NITROGEN (N) RATES ESTIMATED FROM SATELLITE IMAGERY (WHITE CO.)

Daniel Quinn: Department of Agronomy, Purdue University
Ana Morales-Ona: Department of Agronomy, Purdue University

Study Location: White County, Indiana | **Field ID:** County Line
Soil Type: Pella silty clay loam, Mundelein silt loam, and Foresman silt loam
Planting Date: May 3, 2023 | **Harvest Date:** September 29, 2023
Corn Hybrid: DKC62-94 | **Corn Seeding Rate:** 35,500 seeds/ac
Farmer's normal N rate (FNR): 160 (w/o starter N) | **Extra N** (herbicide): 34 | **Total N:** 194 lbs N per acre
Previous Crop: Soybean | **Tillage:** Conventional

RESEARCH TRIAL OVERVIEW:

This NRCS (Natural Resources Conservation Service) funded study examines the feasibility of using satellite imagery to determine maize N status and mid-season optimum N fertilizer rates. Five N fertilizer treatments were established and applied after planting (V2 growth stage) based on the percentage of the farmer's N rate (FNR): (1) 40%, (2) 60%, (3) 80%, (4) 100%, and (5) 120% of the FNR. Plots were 60 feet wide (24, 30-inch corn rows) by the length of the field. Treatments 80, 100, and 120% FNR were replicated two times in a randomized complete block design. Treatments 40 and 60% FNR were replicated three times and placed adjacent to the side of the blocks with the 80, 100, and 120% FNR treatments. A reduced number of replications were established for the higher N rate treatments (80, 100, and 120% FNR) with the aim to reduce overall total N rate applied. Plots were further delineated into shorter sections, "subblocks" equal to the plot width by 300 ft long. Subblocks representing the range of all N rates were considered as a "block." At growth stage V7–V8, variable-rate N fertilizer prescriptions were developed through identification of the agronomic optimum N fertilizer rate (AONR) of each block based on NDVI from satellite (PlanetScope multispectral images, 3-m resolution). A second sidedress N was applied in the form of UAN (28%) in the areas corresponding to the treatments of 40, 60, and 80% FNR. The field was harvested with a commercial combine and adjusted to 15.5% moisture for yield analysis.

RESULTS:

TABLE 37. *Nitrogen (pounds per acre) applications and mean grain yield (bushels per acre) per treatment. White County, IN 2023.*

TREATMENT	EXTRA N*	SIDEDRESS 1	SIDEDRESS 2	TOTAL	GRAIN YIELD
		----------			---- bu/ac ----
		-- lbs N/ac			

40% FNR sidedress 1 + sidedress 2	34	66	78	172	216 a[†]
60% FNR sidedress 1 + sidedress 2	34	95	49	178	210 b
80% FNR sidedress 1 + sidedress 2	34	127	11	173	207 b
100% FNR sidedress	34	157	0	191	210 b
120% FNR sidedress	34	191	0	225	220 a
P–value					<0.001

* Extra N (UAN 28 %) applied with herbicide; 1st sidedress (anhydrous NH_3): May 19 ~V2; 2nd sidedress (UAN 28%): June 19 ~V7
[†] Mean values that do not contain the same corresponding letter are determined statistically different ($P < 0.1$).

SUMMARY (TAKE-HOME POINTS):

- Total N applied across the treatments ranged from 172 to 225 lbs N/ac, with 191 lbs N/ac being the farmer's normal total N rate applied (FNR; Table 37).
- For the treatments that received sidedress N application (40, 60, and 80% FNR + sidedress), total N applied ranged from 172 to 178 lbs N/ac. However, yield response to the 40% FNR + sidedress treatment was higher (216 bu/ac) compared to the other two sidedress treatments (210 and 207 bu/ac).
- Across all treatments examined, the 40% FNR + sidedress and 120% FNR resulted in the highest yield observed.

CORN YIELD RESPONSE TO IN-SEASON NITROGEN (N) RATES ESTIMATED FROM SATELLITE IMAGERY (WHITE CO.)

Daniel Quinn: Department of Agronomy, Purdue University
Ana Morales-Ona: Department of Agronomy, Purdue University

Study Location: White County, Indiana | **Field ID:** Don209
Soil Type: Mundelein silt loam and Pella silty clay loam
Planting Date: May 4, 2023 | **Harvest Date:** October 10, 2023
Corn Hybrid: DKC62-94 | **Corn Seeding Rate:** 35,500 seeds/ac
Farmer's normal N rate (FNR): 180 (w/o starter N) | **Extra N** (herbicide): 34 | **Total N:** 214 lbs N per acre
Previous Crop: Corn | **Tillage:** Conventional

RESEARCH TRIAL OVERVIEW:

This NRCS (Natural Resources Conservation Service) funded study examines the feasibility of using satellite imagery to determine maize N status and mid-season optimum N fertilizer rates. Five N fertilizer treatments were established and applied after planting (V2 growth stage) based on the percentage of the farmer's N rate (FNR): (1) 40%, (2) 60%, (3) 80%, (4) 100%, and (5) 120% of the FNR. Plots were 60 feet wide (24, 30-inch corn rows) by the length of the field. Treatments 80, 100, and 120% FNR were replicated two times in a randomized complete block design. Treatments 40 and 60% FNR were replicated three times and placed adjacent to the side of the blocks with the 80, 100, and 120% FNR treatments. A reduced number of replications were established for the higher N rate treatments (80, 100, and 120% FNR) with the aim to reduce overall total N rate applied. Plots were further delineated into shorter sections, "subblocks" equal to the plot width by 300 ft long. Subblocks representing the range of all N rates were considered as a "block." At growth stage V7–V8, variable-rate N fertilizer prescriptions were developed through identification of the agronomic optimum N fertilizer rate (AONR) of each block based on NDVI from satellite (PlanetScope multispectral images, 3-m resolution). A second sidedress N was applied in the form of UAN (28%) in the areas corresponding to the treatments of 40, 60, and 80% FNR. The field was harvested with a commercial combine and adjusted to 15.5% moisture for yield analysis.

RESULTS:

TABLE 38. *Nitrogen (pounds per acre) applications and mean grain yield (bushels per acre) per treatment. White County, IN 2023.*

TREATMENT	EXTRA N*	SIDEDRESS 1	SIDEDRESS 2	TOTAL	GRAIN YIELD
		------------ lbs N/ac -----------			---- bu/ac ----
40% FNR sidedress 1 + sidedress 2	34	71	76	182	186 c[†]
60% FNR sidedress 1 + sidedress 2	34	107	41	182	187 bc
80% FNR sidedress 1 + sidedress 2	34	142	6	182	191 b
100% FNR sidedress	34	180	0	214	198 a
120% FNR sidedress	34	216	0	250	187 bc
P-value					<0.001

* Extra N (UAN28 %) applied with herbicide; 1st sidedress (anhydrous NH_3): May 20 ~V2; 2nd sidedress (UAN 28%): June 17 ~V7
[†] Mean values that do not contain the same corresponding letter are determined statistically different ($P < 0.1$).

SUMMARY (TAKE-HOME POINTS):

- Total N applied across the treatments ranged from 182 to 250 lbs N/ac, with 213 lbs N/ac being the farmer's normal total N rate applied (FNR; Table 38).
- For the treatments that received sidedress N application (40, 60, and 80% FNR + sidedress), total N applied was 182 lbs N/ac. Yield response was not statistically different for treatments 40 and 60% FNR + sidedress, but it was for treatments 40 and 80% FNR + sidedress.
- Across all treatments examined, the 100% FNR treatment resulted in the highest yields observed.

CORN YIELD RESPONSE TO IN-SEASON NITROGEN (N) RATES ESTIMATED FROM SATELLITE IMAGERY (WHITE CO.)

Daniel Quinn: Department of Agronomy, Purdue University
Ana Morales-Ona: Department of Agronomy, Purdue University

Study Location: White County, Indiana | **Field ID:** Griner
Soil Type: Mundelein silt loam and Pella silty clay loam
Planting Date: May 2, 2023 | **Harvest Date:** October 10, 2023
Corn Hybrid: DKC62-94 | **Corn Seeding Rate:** 35,500 seeds/ac
Farmer's normal N rate (FNR): 180 (w/o starter N) | **Extra N** (herbicide): 34 | **Total N:** 214 lbs N per acre
Previous Crop: Corn | **Tillage:** Conventional

RESEARCH TRIAL OVERVIEW:

This NRCS (Natural Resources Conservation Service) funded study examines the feasibility of using satellite imagery to determine maize N status and mid-season optimum N fertilizer rates. Five N fertilizer treatments were established and applied after planting (V_2 growth stage) based on the percentage of the farmer's N rate (FNR): (1) 40%, (2) 60%, (3) 80%, (4) 100%, and (5) 120% of the FNR. Plots were 60 feet wide (24, 30-inch corn rows) by the length of the field. Treatments 80, 100, and 120% FNR were replicated two times in a randomized complete block design. Treatments 40 and 60% FNR were replicated three times and placed adjacent to the side of the blocks with the 80, 100, and 120% FNR treatments. A reduced number of replications were established for the higher N rate treatments (80, 100, and 120% FNR) with the aim to reduce overall total N rate applied. Plots were further delineated into shorter sections, "subblocks" equal to the plot width by 300 ft long. Subblocks representing the range of all N rates were considered as a "block." At growth stage V_7–V_8, variable-rate N fertilizer prescriptions were developed through identification of the agronomic optimum N fertilizer rate (AONR) of each block based on NDVI from satellite (PlanetScope multispectral images, 3-m resolution). A second sidedress N was applied in the form of UAN (28%) in the areas corresponding to the treatments of 40, 60, and 80% FNR. The field was harvested with a commercial combine and adjusted to 15.5% moisture for yield analysis.

RESULTS:

TABLE 39. *Nitrogen (pounds per acre) applications and mean grain yield (bushels per acre) per treatment. White County, IN 2023.*

TREATMENT	EXTRA N*	SIDEDRESS 1	SIDEDRESS 2	TOTAL	GRAIN YIELD
		------------ lbs N/ac -----------			----bu/ac----
40% FNR sidedress 1 + sidedress 2	34	73	67	172	200 bc[†]
60% FNR sidedress 1 + sidedress 2	34	109	31	174	197 c
80% FNR sidedress 1 + sidedress 2	34	144	2	180	204 a
100% FNR sidedress	34	178	0	212	198 c
120% FNR sidedress	34	215	0	249	204 a
P-value					0.10

* Extra N (UAN 28 %) applied with herbicide; 1st sidedress (anhydrous NH_3): May 19 ~V2; 2nd sidedress (UAN 28%): June 17 ~V7

[†] Mean values that do not contain the same corresponding letter are determined statistically different ($P < 0.1$).

SUMMARY (TAKE-HOME POINTS):

- Total N applied across the treatments ranged from 172 to 249 lbs N/ac, with 210 lbs N/ac being the farmer's normal total N rate applied (FNR; Table 39).

- For the treatments that received sidedress N application (40, 60, and 80% FNR + sidedress), total N applied ranged from 172 to 180 lbs N/ac. However, yield response to the 80% FNR + sidedress treatment was greater (204 bu/ac) compared to the other two sidedress treatments (200 and 197 bu/ac).

- Across all treatments examined, the 80% FNR + sidedress and the 120% FNR sidedress resulted in the highest yields observed.

CORN YIELD RESPONSE TO IN-SEASON NITROGEN (N) RATES ESTIMATED FROM SATELLITE IMAGERY (MARSHALL CO.)

Daniel Quinn: Department of Agronomy, Purdue University
Ana Morales-Ona: Department of Agronomy, Purdue University

Study Location: Marshall County, Indiana | **Field ID:** EV
Soil Type: Crosier loam and Riddles-Oshtemo fine sandy loam
Planting Date: May 12, 2023 | **Harvest Date:** October 23, 2023
Corn Hybrid: DKC59-82 | **Corn Seeding Rate:** 32,000 seeds/ac
Farmer's normal N rate (FNR): 150 (w/o starter N) | **Starter N:** 30 | **Total N:** 180 lbs N per acre
Previous Crop: Soybean | **Tillage:** Conventional

RESEARCH TRIAL OVERVIEW:

This NRCS (Natural Resources Conservation Service) funded study examines the feasibility of using satellite imagery to determine maize N status and mid-season optimum N fertilizer rates. Five N fertilizer treatments were established and applied before planting based on the percentage of the farmer's N rate (FNR): (1) 40%, (2) 60%, (3) 80%, (4) 100%, and (5) 120% of the FNR. An additional treatment equivalent to the farmer's practice (starter and sidedress N only, no preplant N) was established too. Plots were 40 feet wide (16, 30-inch corn rows) by the length of the field. Treatments FP, 80, 100, and 120% FNR were replicated four times in a randomized complete block design. Treatments 40 and 60% FNR were replicated seven times and placed adjacent to each side of the blocks with the FP, 80, 100, and 120% FNR treatments. A reduced number of replications were established for the higher N rate treatments (80, 100, and 120% FNR) with the aim to reduce overall total N rate applied. Plots were further delineated into shorter sections, "subblocks" equal to the plot width by 200 ft long. Subblocks representing the range of all N rates were considered as a "block." At growth stage V6, variable-rate N fertilizer prescriptions were developed through identification of the agronomic optimum N fertilizer rate (AONR) of each block based on NDVI from satellite (PlanetScope multispectral images, 3-m resolution). Sidedress N was applied in the form of UAN (28%) in the areas corresponding to the treatments of 40, 60, and 80% FNR. The field was harvested with a commercial combine and adjusted to 15.5% moisture for yield analysis.

RESULTS:

TABLE 40. *Nitrogen (pounds per acre) applications and mean grain yield (bushels per acre) per treatment. Marshall County, IN 2023.*

TREATMENT	PREPLANT*	STARTER	SIDEDRESS	TOTAL	GRAIN YIELD
	------------ lbs N/ac -----------				---- bu/ac ----
Farmer's practice	0	30	150	180	265 a[†]
40% FNR sidedress	60	30	68	158	261 a
60% FNR sidedress	90	30	38	158	265 a
80% FNR sidedress	120	30	13	163	265 a
100% FNR	150	30	0	180	263 a
120% FNR	180	30	0	210	264 a
P–value					0.078

* Preplant (UAN28 %): April 20; Starter (liquid mix): May 12; Sidedress (UAN 28%): June 10 ~V6

[†] Mean values that do not contain the same corresponding letter are determined statistically different (*P* < 0.1).

SUMMARY (TAKE-HOME POINTS):

- Total N applied across the treatments ranged from 158 to 210 lbs N/ac, with 180 lbs N/ac being the farmer's normal total N rate applied (FNR; Table 1).
- Across all treatments examined, yield response was not statistically different regardless of the different total N rate applied.
- For the treatments that received sidedress N application (40, 60, and 80% FNR + sidedress), total N applied ranged from 158 to 163 lbs N/ac and yield response was not statistically different.

CORN YIELD RESPONSE TO IN-SEASON NITROGEN (N) RATES ESTIMATED FROM SATELLITE IMAGERY (DUBOIS CO.)

Daniel Quinn: Department of Agronomy, Purdue University
Ana Morales-Ona: Department of Agronomy, Purdue University

Study Location: Dubois County, Indiana | **Field ID:** DU
Soil Type: Dubois silt loam, Otwell silt loam, Peoga silt loam
Planting Date: May 3, 2023 | **Harvest Date:** October 11, 2023
Corn Hybrid: P1718AML | **Corn Seeding Rate:** 32,500 seeds/ac
Farmer's normal N rate (FNR): 185 (w/o starter N) | **Starter N:** 52 | **Total N:** 237 lbs N per acre
Previous Crop: Wheat | **Tillage:** No-till

RESEARCH TRIAL OVERVIEW:

This NRCS (Natural Resources Conservation Service) funded study examines the feasibility of using satellite imagery to determine maize N status and mid-season optimum N fertilizer rates. Five N fertilizer treatments were established and applied before planting based on the percentage of the farmer's N rate (FNR): (1) 40%, (2) 60%, (3) 80%, (4) 100%, and (5) 120% of the FNR. Plots were 80 feet wide (32, 30-inch corn rows) by the length of the field. Treatments 40, 60, 80, and 100% FNR were replicated three times in a randomized complete block design. Treatment 120% FNR was replicated two times and placed between the blocks with the 40, 60, 80, and 100% FNR treatments. Plots were further delineated into shorter sections, "subblocks" equal to the plot width by 200 ft long. Subblocks representing the range of all N rates were considered as a "block." At growth stage V6, variable-rate N fertilizer prescriptions were developed through identification of the agronomic optimum N fertilizer rate (AONR) of each block based on NDVI from satellite (PlanetScope multispectral images, 3-m resolution). Sidedress N was applied in the form of urea in the areas corresponding to the treatments of 40, 60, and 80% FNR. The field was harvested with a commercial combine and adjusted to 15.5% moisture for yield analysis.

RESULTS:

TABLE 41. *Nitrogen (pounds per acre) applications and mean grain yield (bushels per acre) per treatment. Dubois County, IN 2023.*

TREATMENT	PREPLANT*	STARTER	SIDEDRESS	TOTAL	GRAIN YIELD
	------------ lbs N/ac -----------				----bu/ac----
40% FNR sidedress	80	53	62	189	247 c[†]
60% FNR sidedress	110	53	33	197	251 bc
80% FNR sidedress	145	53	0	198	257 ab
100% FNR	180	53	0	233	261 a
120% FNR	215	53	0	267	258 ab
P–value					<0.001

* Preplant (anhydrous NH$_3$): April 13; Starter (liquid mix): May 3; Sidedress (urea): June 10 ~V6

[†] Mean values that do not contain the same corresponding letter are determined statistically different ($P < 0.1$).

SUMMARY (TAKE-HOME POINTS):

- Total N applied across the treatments ranged from 189 to 267 lbs N/ac, with 233 lbs N/ac being the farmer's normal total N rate applied (FNR; Table 1).
- For the treatments that received sidedress N application (40, 60, and 80% FNR + sidedress), total N applied ranged from 189 to 198 lbs N/ac. Yield response to the 40 and 60% FNR + sidedress treatment was lower (247 and 251 bu/ac respectively) compared to the other treatments, which ranged from 257 to 261 bu/acre.
- Across all treatments examined, the 80% FNR, 100% FNR, and 120% FNR resulted in the highest yield observed. However, statistically yield was not different from the yield observed with the 80% FNR + sidedress treatment or the 120% FNR treat5ment.

INTERESTED IN PARTICIPATING IN ON-FARM RESEARCH?

Interested in working with Purdue University to address management questions and improve your operation through on-farm research? Both the Purdue Corn Agronomy Team and the Purdue On The Farm Program continue to look for on-farm cooperators for participation and assistance with on-farm research trials. In addition, we will work closely with you to answer specific questions both we and you may have specific to your own operation. Information and data collected are shared directly to each cooperator every step of the way. For more information, please reach out directly below:

Daniel Quinn, PhD
Extension Corn Specialist
Purdue University
Email: djquinn@purdue.edu
Office: 765-494-5314.

ABOUT THE AUTHOR

DANIEL QUINN is an assistant professor of agronomy and an extension corn specialist at Purdue University. His research and extension program focuses on improving corn production systems in the Midwest through large-scale and small-plot field trials. Quinn's key areas of study include yield physiology, agronomic intensification, precision technologies, nutrient management, and cover crops, with an emphasis on enhancing profitability, productivity, and sustainability in corn-based agriculture.